EMPower
extending mathematical power

AF882274

# Many Points Make a Point

## Data and Graphs

**TEACHER BOOK**

TERC
Because math and science build futures

Mary Jane Schmitt, Myriam Steinback, Tricia Donovan, and Martha Merson

TERC 2067 Massachusetts Avenue
Cambridge, Massachusetts 02140

**_EMPower_ Authors / Research and Development Team**
Mary Jane Schmitt, Tricia Donovan, Myriam Steinback, and Martha Merson

**Contributors and Reviewers**
Donna Curry, Aren Lew, Marlene Kliman, Pam Meader, Connie Rivera, Heidi Schuler, and Sherry Soares

**Technical Team**
Production and Design Team: Juania Ashley, Valerie Martin, and Sherry Soares
Copyeditor: Jill Pellarin

**Evaluation Team**
Brett Consulting Group:
Belle Brett
Marilyn Matzko

*EMPower*™ was developed at TERC in Cambridge, Massachusetts. This material is based upon work supported by the National Science Foundation under award number ESI-9911410 and by the Education Research Collaborative at TERC. Any opinions, findings, and conclusions or recommendations expressed in this publication are those of the authors and do not necessarily reflect the views of the National Science Foundation.

TERC is a nonprofit made up of teams of math and science education and research experts dedicated to innovation and creative problem solving. At the frontier of theory and practice, TERC's work encompasses research, content and curriculum development, technology innovation, professional development, and program evaluation. TERC has a passion for social justice and strives to create level playing fields for all learners, reaching more than three million students every year.



https://www.terc.edu/empower

Printed in the United States of America

1 2 3 4 5 6 7 8 9 10

ISBN 978-1-7373997-8-0



Send all *EMPower* content inquiries to:

Adult Numeracy Center at TERC
empower@terc.edu

Adult Numeracy Center

## Advisory Board

Donna Curry, Equipped for the Future, National Center
Solange Farina, Math Exchange Group, New York City
Linda Kime, Mathematics Association of America and
University of Massachusetts, Boston
Ron Kindig, Los Angeles Unified School District (retired)
Kenn Pendleton, General Education Development (GED) Testing Service
Ujwala Samant, National Center for the Study of Adult Learning and Literacy
William Thomas, American Mathematics Association of Two-Year Colleges
(AMAYTC) and University of Toledo

## Consultants and Reviewers

Scott W. Beckett, Jacksonville State University
Sherry Fraser, Interactive Mathematics Program
Arthur G. Fruhling, Yuba Community College
Nancy R. Johnson, Manatee Community College
Steve Monk, University of Washington
Elizabeth Phillips, Connected Mathematics Program
Mark Rudd, Albuquerque TVI
Karen Sitren, Santa Fe Community College
Cornelia Tierney, TERC
Kevin Wheeler, Three Rivers Community College

## EMPower Teacher Participants

**Arizona**

Gila River Indian Community 21st Century Community Learning Center, Gila River
Andrea Parrella

**Illinois**

Adult Education Center, Rock Valley College, Rockford
Phyllis Flanagan, Melinda Harrison

Comprehensive Community Solutions, Inc., YouthBuild, Rockford
Ryan Boyce

Township High School District 214 Community Education, Arlington Heights
Kathy Conrad

**Massachusetts**

Action for Boston Community Development, Inc., LearningWorks, Boston
Kelly Qualman

Adult Learning Center at Middlesex Community College, Bedford
Roberta Froelich

Brockton Adult Learning Center, Brockton
Marilyn Moses

Community Learning Center, Cambridge
Nellie Dedmon, Sylvia Lotspeich Greene, Linda Huntington

Dimock Commuity Health Center, Roxbury
Martha Gray

Harvard Bridge to Learning and Literacy, Cambridge
Judy Hikes, Carol Kolenik

Holyoke Adult Learning Opportunities (HALO) Center, Holyoke
Kelly Martin, Glenn Yarnell

Hyde Square Task Force, Jamaica Plain
Nora O'Connor

Jewish Vocational Services, Boston
Susan Arase, Terry Lerner, Jeff Snyder

Lowell Adult Education Program, Lowell
Barbara Goodridge

Mount Wachusett Community College, Gardner
Patricia Vorfeld

Notre Dame Education Center, South Boston
Esther D. Leonelli

Pioneer Valley Adult Education Center, Northampton
Olu Odusina

Project Place, Boston
Sharon Carey, Maia Hendrickson

Quinsigamond Community College Adult Basic Education Program, Worcester
Cathy Coleman

Read/Write/Now, Springfield
Lee E. Boone, Susanne Campagna, Michelle Faith Brown, Lucille Fandel

Somerville Center for Adult Learning Experiences (SCALE), Somerville
Tom Glannon

X-Cel Adult Education, Dorchester
Don Sands

**Maine**

Bailey Evening School, Bath
Pam Bessey

Gardiner Adult and Community Education Program
Diann Bailey

Windham Adult Education, Windham
Eva Giles

## EMPower Teacher Participants *(continued)*

**New Jersey**

New Brunswick Public Schools—Adult Learning Center, New Brunswick
Jacqueline Arkoe, Phyllis Boulanger, Sue Helfand, Karen C. Pickering

**New York**

Adult Learning Center, La Guardia Community College, Long Island City
Norma Andrade, Mark Trushkowsky

BEGIN Managed Programs, New York
Charles Brover, Solange Farina, Santiago Perez, Elizabeth Reddin

Borough of Manhattan Community College/CUNY Adult Basic Education Program, Center for Continuing Education and Workforce Development, New York
Elliot Fink, Mark Lance

City University of New York, Continuing Education, New York
Lisa Simon

Lehman Adult Learning Center, Bronx
Deidre Freeman

SafeSpace-LifeSkills, New York
Jonna Rao

The Adult Learning Center, College of Staten Island, Staten Island
Karen Johnsen

**Pennsylvania**

Lancaster-Lebanon Intermediate Unit 13/Career Link, Lancaster
Margaret Giordano, Barbara Tyndall

**Rhode Island**

Dorcas Place, Parent Literacy Center, Providence
Shannon Dolan, Jerelyn Thomas

**Tennessee**

Hawkins County Adult Education Program, Rogersville
Lisa Mullins

## Acknowledgments

Many friends and family members supported the teachers' and *EMPower's* efforts. Thank you Denise Deagan, Judith Diamond, Cara Dimattia, Ellen McDevitt, Sandy Strunk, and the board of the Adult Numeracy Network. Roberta Froelich, Brad Hamilton,Michael Hanish, David Hayes, Judy Hikes, Alisa Izumi, Esther D. Leonelli, Myra Love, Lambrina Mileva, Luz Rivas, Johanna Schmitt, Rachael Stark, Jonathan Steinback, and Sean Sutherland all made unique and timely contributions to the project.

We appreciate the encouragement and advice from John (Spud) Bradley, *EMPower's* program officer, and Gerhard Salinger of the National Science Foundation.

We are indebted to every adult student who participated in the piloting of *EMPower*. Their honest feedback and suggestions for what worked and what did not work were invaluable. We could not have completed the curriculum without them or their teachers.

# Contents

# Introduction to the EMPower Series

## Background

*Extending Mathematical Power (*EMPower*)* integrates recent mathematics education reform into the field of education for adults and out-of-school youth. *EMPower* was designed especially for those students who return for a second chance at education by enrolling in remedial and adult basic education programs, high school equivalency programs, and developmental programs at community colleges. However, the series is appropriate for a variety of other settings as well, such as high schools, workplaces, and parent and paraprofessional education programs. *EMPower* builds interest and competency in mathematical problem solving and communication.

Over the course of four years (2000–2004), a collaboration of teachers and researchers with expertise in adult numeracy education and K–12 mathematics reform developed and piloted eight contextualized units. These units are organized around four central topics: number and operation sense; patterns, functions, and relations; geometry and measurement; and data and graphs. The *EMPower* program serves as a model for a cohesive mathematics series that offers content consistent with the *Principles and Standards for School Mathematics* (NCTM, 2000), as well as frameworks that are adult-focused, such as the *Equipped for the Future Content Standards* (Stein, 2000), the *Massachusetts ABE Curriculum Frameworks for Mathematics and Numeracy* (Massachusetts Department of Education, 2001), and the *Adult Numeracy Network's Framework for Adult Numeracy Standards* (Curry, Schmitt, & Waldron, 1995). The program fosters a pedagogy of learning for understanding; it embeds teacher support and is transformative, yet realistic, for multilevel classrooms.

*EMPower* challenges students and teachers to consistently extend their ideas of what it means to do math. The series focuses on mathematical reasoning, communication, and problem solving with a variety of approaches and strategies, not just rote memorization and symbol manipulation. The program fosters a learning community in which students are encouraged to expand their understanding of mathematics through open-ended investigations, working collaboratively, sharing ideas, and discovering multiple ways for solving problems. The goal of *EMPower* is to help people build experience managing the mathematical demands of various life situations, such as finances and commerce, interpretation of news stories, and leisure activities, and to connect those experiences to mathematical principles.

## A Focus on Mathematical Content

The *EMPower* series supports students' and teachers' growth by directing attention to significant mathematical understandings.

### *EMPower* emphasizes:

- Data analysis, geometry and measurement, algebra, and number and operations at all student levels.
- Reliance on benchmark numbers—such as powers and multiples of 10, common fractions, and their decimal and percent equivalents—for making mental calculations.
- Early use of calculators to support computation.
- Development of reasoning on proportion and parts of quantities before consideration of formal operations with rational numbers.
- Making decisions about data where students generate, as well as interpret, graphical representations.
- Geometry and measurement based on opportunities to see and touch in developing an understanding of spatial relationships and formulas.
- Leading with patterns and relationships in contextual situations and the representations of these situations with diagrams, tables, graphs, verbal rules, and symbolic notation to develop algebraic competence.

## A Focus on Pedagogy

Mathematics is meaningful within a social context. While mathematical truths are universal, the meaning and relevance of numbers changes according to the setting and culture. Therefore, the *EMPower* pedagogy is focused on sets of connected activities that require communication and discourse.

*EMPower* asks students to

- Work collaboratively with others on open-ended investigations;
- Share strategies orally and in writing; and
- Justify answers in multiple ways.

Key features of activities provide

- Clear mathematical goals
- Contexts that are engaging and useful for young people and adults
- Opportunities to strengthen mathematical language and communication skills
- Various ways of entering and solving problems
- Puzzles that draw students into problems and motivate them to seek a solution.

# Overview of EMPower Units

## *Features of the Teacher Book*

FACILITATING LESSON **1**

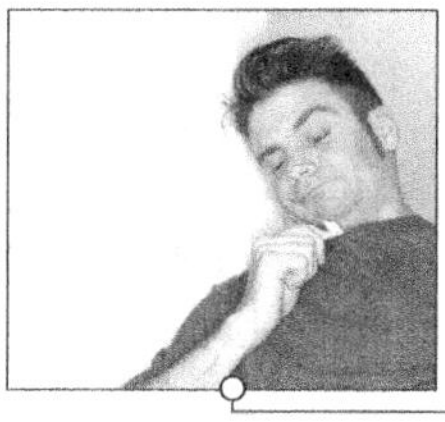

### Countries in Our Closets

***Where were your clothes made?***

#### Synopsis

This is the first of a set of lessons on categorical data in which students collect, examine, sort, and organize their own data, and then compare their data to another data set.

1. Students bring to the class data on the different geographic locations where their clothes were made.
2. The class combines its data into one frequency graph, and students make several statements about it.
3. Working in pairs, students compare their class graph to data from another class.
4. Pairs of students each create a new frequency graph, organizing the class data by continent. Students notice which continent is most prevalent.
5. Discussion centers on comparing country and continent graphs, closing with the question How does the story change when the graph's categories change?

#### Objectives

- Use a frequency graph to organize data
- Identify the story that the data tell
- Compare data from various samples
- Change the categories and articulate the change in the story

EMPower™ Facilitating Lesson 1: Countries in Our Closets 15

Each lesson opens with a **synopsis** of the lesson activities.

**Objectives** summarize the skills students should gain from the lesson.

#### Materials/Prep

- Grid paper
- Paper strips for posting statements
- Post-it® Notes
- Rulers

As an aid for determining in which continent the countries are located, have on hand a world atlas such as *The New Children's Illustrated Atlas of the World,* by Keith Lye (Philadelphia: Courage Press, 1999). For a listing of countries in Asia, see *Resource: Countries in Asia, Blackline Master 4.*

Make sure you have assigned *Countries in Our Closets* (*Student Book,* p. 7). Collect and bring in about 20 items of clothing in case your students have not collected their own data. (See *Facilitation,* p. 25, for how to handle this.)

Prepare a transparency or make a copy of the graph "Countries in Our Closets, Springfield, MA" (*Student Book,* p. 10).

#### Opening Discussion

Ask students:

**Are clothes made in this city? In this country? Why or why not?**

**Why would anyone care where his or her clothes were made?**

Acknowledge that we should know something about where our clothes come from, given how clothing is worn next to our skin. There are also tremendous economic and political issues at stake as manufacturing jobs are outsourced outside of the United States.

#### Heads Up!

If students do not bring in data, see *Facilitation,* p. 25.

**You have looked in your closets, checked the tags on some of your clothes, and collected information. What did you learn?**

**Which country do most of the clothes you checked come from?**

Post the country names in a place where you can easily refer to them later.

Introduce the terms **data set** and **mode**. Each student's list is a data set, and each data set could have a mode—the country name that occurs most frequently.

*Vocabulary note:* If students are searching for words to express their thoughts, offer terms and concise definitions. Students can keep track of these in their student books. For example, **data** are the information collected. The data that are collected make up the **sample.** The whole group from which information could be collected is the **population.** See *Lesson 1 Commentary,* p. 24, for more information.

16 Many Points Make a Point: Data and Graphs EMPower™

The **Materials/Prep** feature tells teachers what materials they will need to prepare for the day's activities.

The **Opening Discussion** gives teachers ideas for engaging students at the beginning of the lesson.

**Heads Up!** alerts teachers to helpful hints and things to look for in facilitating the lesson.

Dialog balloons and bold text suggest statements and questions teachers can use to spark and guide class discussion.

A **Summary Discussion** guides the conclusion of the lesson.

**Practice, Extension,** and **Test Practice** sections list and describe choices for additional work.

**Looking Closely** draws the teacher's attention to the goals and suggests teaching strategies.

**Math Background** helps teachers deepen their understanding of mathematical concepts.

The **What to Look For** grid allows teachers to track students' progress and think ahead about next steps.

---

### Summary Discussion

DATA

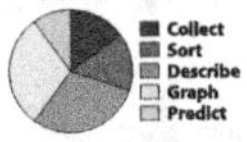


Finish the lesson by posing the following questions. (Students may not have ready answers at this point, but the questions will be a thread through the rest of the lessons.)

**What stories can be told from this set of data (or what questions can be asked and answered)?**

**How does organizing the information into different categories impact how the data may be interpreted?**

**Who would be curious about what the people in this class eat?**

Direct student to the *Reflections* questions (*Student Book*, p. 145).

### Practice

*Take a Sample*, p. 24
For practice choosing a representative sample.

*Thirsty* and *Ouch!*, pp. 26 and 28
For practice organizing information and making statements.

### Extension

*Friends and Family Drink Too*, p. 30

*Data Collection*, p. 31
Students begin recording what they eat for a project they will do in a later lesson. For now, they just write down all the foods they eat throughout the day.

### Test Practice

*Test Practice*, p. 32

### Looking Closely

Observe whether students are able to

**Organize data into consistent categories**

Do students group similar foods together? Make sure students are moving index cards around and that they are writing down the category names. If students seem stuck, have them ask you specific questions.

If you see foods categorized incorrectly, ask a question such as "Do all group members agree with the categories?" Encourage students to think, question, and defend their reasoning. If students use overlapping categories such as "cereal" and "grain," present an example close to their own experience. Say, for instance, "It is

EMPower™ Facilitating Lesson 2: Most of Us Eat... 33

---

**LESSON 1 COMMENTARY**

### Rationale

The lesson takes students through the first steps of displaying data: collecting and organizing the information. As you work through this lesson, you and your students will notice that each time you categorize and recategorize the data set, you tell a different story.

### Math Background

Data are information. This information may be numerical—e.g., salaries, test scores, heights, age, weight—as well as categorical—e.g., countries, foods we eat, types of product defects.

Most data are reported using statistics based on fractions and percents—one-third of pregnant mothers, half of native plants, etc. Commonly used, or benchmark, fractions and percents are often invoked to influence others to make decisions. When the headlines shout, "One-half of all pregnant women …," readers tend to form opinions based on the data. What readers often neglect to consider is the size of the sample, where the data were collected, and how the data were organized.

The media generally report data that are based on a sample. It is rare for any group to have the time or resources to ask every person in the population to contribute information. The idea of sampling is to study a part in order to gain information about the whole. The sample—its size and characteristics—influences the data and the conclusions of the study.

A representative sample by definition includes representation across the population being surveyed. For example, a representative sample of the community would include individuals from across town, not just one neighborhood. Likewise, a representative sample across America would include individuals from many different states and regions, not just one or two. The size of the sample is important when inferences are made. For example, 30 samples could be sufficient for a population of several hundreds. The deciding factor for determining sample size is how confident you want to be about the inference.

It is not the concern of this unit to determine appropriate sample size, but it is important to note that statisticians use formulas derived from repeated surveys to decide sample size. Regardless, small samples are used in many studies. Factors … might lead … outcome.

…k critically … or into …ield …*esson 2.*

EMPower™

---

EMPower™

| WHAT TO LOOK FOR IN *LESSON 1* | WHO STANDS OUT? (LIST STUDENTS' INITIALS) STRONG | ADEQUATE | NEEDS WORK | NOTES FOR NEXT STEPS |
|---|---|---|---|---|
| Concept Development<br>• Records all data<br>• Organizes data<br>• Recognizes key features of frequency graph<br>• Compares data<br>• Changes categories and notes change in story | | | | |
| Expressive Capacity<br>• Uses precise language to describe the frequency graphs<br>• Contributes to the discussion about the way changing categories affects the story the data tell | | | | |
| Use of Tools<br>• Uses graph paper<br>• Uses atlas or other reference | | | | |
| Background Knowledge<br>• Draws on understanding of geography, relative size, and number sense for comparisons | | | | |

Facilitating Lesson 1: Countries in Our Closets 23

### Context

Some students may know about *maquiladoras* in Mexican border towns, where women make clothes for very little money and with no benefits or environmental Occupational Safety and Health Administration (OSHA) workplace protections. CorpWatch (www.corpwatch.org) is one source for information on *maquiladoras*.

### Facilitation

If students do not bring in data, or if their sample is too small, skip the second part of the *Opening Discussion*. Have available a pile of 20 clothing articles with labels. First, ask students to predict where the clothes were made. Post the list of their guesses. Note that it will be hard for them to answer this question unless they organize the information on the labels. Then divide up the 20 articles of clothing. Have students write the name of the country for each piece of clothing on a Post-it Note, one country name per note. Ask: "Where are most of our clothes made?" Then continue with the activity.

The authors give ideas for **Making the Lesson Easier** and **Making the Lesson Harder.**

#### Making the Lesson Easier

Frequency graphs lend themselves to comparisons among categories. If students have little fluency stating comparisons, you may choose only to compare size, using terms like "greater," "fewest," or "less than." For students who are encountering data formally for the first time, the notion that collapsing data yields different stories may be difficult. Treat this lightly in the activity, and revisit such questions after students have more experience categorizing and recategorizing data in the homework and in *Lesson 2*.

#### Making the Lesson Harder

If your students can handle benchmark fractions and percents, get them to look critically at the data, including the source and sample size. You might ask:

**If we asked another class what countries are in their closets, what do you think would happen to the categories? What if we asked the entire community?**

**How do you think your data would compare to data from another class of adult students in another community?**

If students struggle with the idea of sample, you might try this: Have them each write their favorite color on a Post-it Note. If you have a small class, ask them to write the color on two Post-it Notes. Place all of the notes in a container. Have someone randomly (eyes closed) choose a few notes from the container and place them across a line to form a frequency graph. Ask the students how they think this sample compares to the actual total number of colors on notes in the container. You can have them do another frequency graph to compare the sample to the actual total.

Facilitating Lesson 1: Countries in Our Closets 25

### LESSON 1 IN ACTION

*Alice articulates the mathematical principle behind compressed data.*

I asked, "How did the change in categories affect what we noticed about the data?"

Alice answered, "Well, we keep losing information."

"How so?"

Patiently, Alice explained that when we started our work, every bit of data was visible. She added that we had lost details initially recorded. "At first, we knew every country in every person's closet and how many pieces of clothing came from that country. Then we combined the data, and we lost track of who had which countries. Then we did it by continent, and we lost track of all the countries."

Alice's realization quickly gained agreement from the rest of the class. After all, just the previous week a classmate had noted, "When you change the amount of data you look at, you find different things."

Sonia added her comment with increased conviction: "It is like politics. Politicians use a graph and tell you this is true, but you look at the graph, and it does not tell you everything."

*Tricia Donovan*
*Pioneer Valley Adult Education Center, Northampton, MA*

In **Lesson in Action,** EMPower teachers share their classroom experiences.

26 Many Points Make a Point: Data and Graphs EMPower™

# Overview of EMPower Units

## *Features of the Student Book*

LESSON **1**

### Countries in Our Closets

***Where were your clothes made?***

***In a world*** where information is at our fingertips, it is important to be able to make sense of it. What does it tell us? Why do we care?

Begin with your own clothes. Where were they made? Find out! You will collect **data** by reading the labels on your clothes. You will organize that data, make statements about it, and note where most of yo[ur clothes] were made.

You will then try organizing the data in a different way. Ho[w] the data in new groups change the story of the countries in [...] You will compare your class data to data of other classes. W[...] be the same?

**On Your Own**

I think the country in which most of my clothes were made [...]

Look at the labels of at least eight items of clothing. Make a [...] track of the information you are collecting. Research questi[on] were your favorite clothes made?

EMPower™ Lesson 1: Countries in [...]

An **introduction** explains the purpose and content of the lesson.

**Activity 1: Organizing the Data**

Look at the data you collected about countries where your clothes were made.

Organize the data.

1. What strikes you about your data?

2. What is the **mode** for your data?

3. What is the mode for the class data?

8 Many Points Make a Point: Data and Graphs EMPower™

**Vocabulary** words are easy to identify in bold type.

The class **activity** comes next. Often students work with a partner or in a small group.

**Practice** pages invite students to use new skills.

### Practice: Categorically Speaking

Sometimes we can organize categories into subgroups of other categories in order to better understand the data. For example, drug stores will often organize their aisles by category—skin care, beauty aids, and so on. Then the items in each of those aisles are also organized. This makes it easier to take inventory and place orders and for the customer to find products.

Can you think of other examples where information is categorized?

Use the chart below *or* create your own chart; draw a picture; make a list; or write about examples you see at home or at work. In the chart below, one example is given.

| Location | What Stuff or Information? | How Is It Organized? |
|---|---|---|
| Drug store | Over-the-counter products they sell | Skin care, Beauty aids, Seasonal, Stationery, First aid, Hair care |
| | | |
| | | |
| | | |

EMPower™ Lesson 1

### Extension: Taking Inventory

Find a drawer or closet filled with many things. Use a frequency graph to show the contents. Start with six categories. Then do another frequency graph to show the same items, but this time use three categories.

For example:

**My Medicine Cabinet**

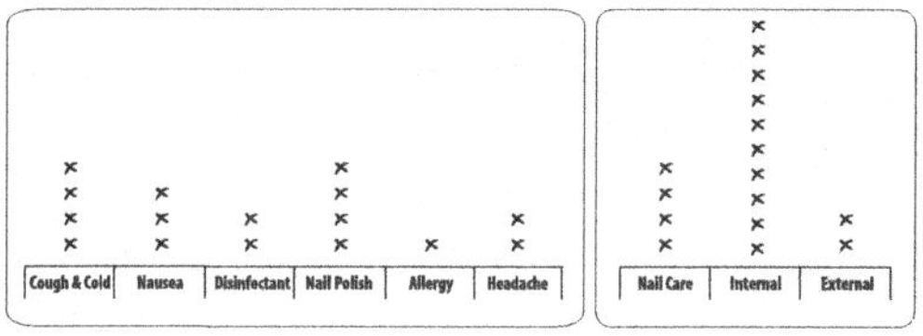


16 Many Points Make a Point: Data and Graphs EMPower™

**Extension** pages present a challenge or a new but related math idea.

Ample space allows students to record most work in their books.

A **Test Practice** that reflects the format of the GED test concludes each lesson.

## Test Practice

1. Freda worked at a printing press. She kept a tally of all the defects she found in the books she was processing. Based on the frequency graph she created, shownbelow, which of the following statements is true?

   (1) There were more defects related to colors than any other defect.

   (2) There were twice as many torn pages as there were warped spines.

   (3) There were more warped spines than colors that bled.

   (4) Half of all the defects were related to colors that bled.

   (5) Half of all the defects were related to pages that were torn.

Book Defects Frequency Graph

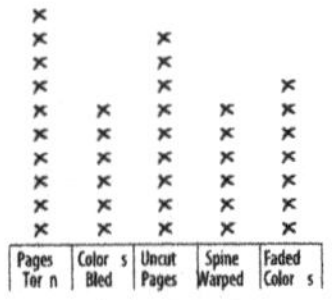


2. Tony, a tour guide, has been keeping a tally of visitors from different states. According to his tally, which of the following is *not* a true statement?

   (1) The fewest number of visitors came from Oregon.

   (2) There were as many visitors from Maine as there were from Georgia.

   (3) There were as many visitors from Georgia as there were from Oregon and Texas.

   (4) There were twice as many visitors from Florida as there were from Rhode Island.

   (5) There were twice as many visitors from Idaho as there were from Maine.

U.S. Tourist Frequency Graph

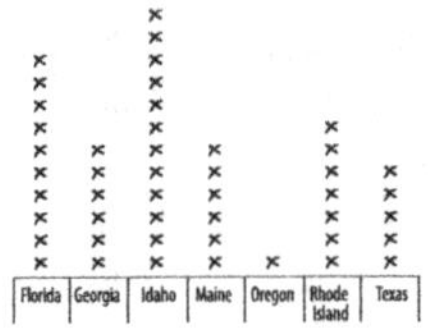


3. Clara is paid a commission on each of the large electronics devices that she sells. She tallied the different items she sold for the month. Based on her tally, what can she tell her boss about her sales?

   (1) She sold twice as many DVD players as she did TVs.

   (2) She sold twice as many DVD players as she did computers.

   (3) She sold more DVD players than she did TVs and computers combined.

   (4) She sold more TVs and VCRs than she did stereos and DVD players combined.

   (5) She sold half as many TVs as she did stereos.

Home Media Frequency Graph

EMPower™

4. Bronson, a forest ranger in the Green Mountains, kept track of the different animals that were reportedly seen in one of the campgrounds during the month of May. Based on his tally, which of the following statements could he tell the media?

   (1) There were twice as many wolves reportedly seen as bears.

   (2) Half of all the reported sightings were wolves.

   (3) There were more bears than wolves reportedly seen.

   (4) There were twice as many moose reportedly seen than there were black bears.

   (5) One-quarter of all reported sightings were brown bears.

North American Wildlife Frequency Graph

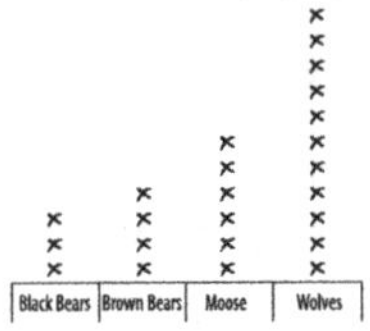


5. According to Global Exchange, some workers in China who make clothing for Disney are paid as little as $0.16 per hour. At this wage, what amount would a worker make for a 40-hour workweek?

   (1) $64.00

   (2) $6.40

   (3) $6.00

   (4) $80.00

   (5) $0.60

6. In 2003 the U.S. Department of Labor (DOL) issued new regulations for overtime pay. The DOL estimated that under the new regulations 1.3 million low-wage workers would become eligible for overtime pay, unless their wages were raised to $425 per week. The DOL estimated that 24.8 percent of those workers were Hispanic and 16.6 percent were African American. What percent were neither Hispanic nor African American?

18 Many Points Make a Point: Data and Graphs EMPower™

## Changing the Culture

The authors have created this series to follow the National Council of Teachers of Mathematics (NCTM) Principles and Standards; however, every teacher who uses *EMPower* faces the challenge of transforming the prevailing culture of his or her math classroom. *EMPower* pilot teachers offer some ideas for facilitating this transition:

- Set the stage. As a class, set ground rules. Explicitly state that this is a space for everyone to learn. As one teacher said, "We are in this together. Share, even if you do not think you are right. Whatever you add will be helpful. It lets us see how you are looking at things."
- Group your students. Match students whose learning styles and background knowledge complement each other. Ask questions, such as, "How did it go to work together?" "How did everyone contribute?"
- Allow wait time. Studies have shown that teachers often wait less than three seconds before asking another question. Students need more time to think.
- Sit down. Watch students before interrupting to help them. Listen for logic and evidence of understanding. Follow the thread of students' thinking to uncover unconventional approaches. During discussions with the whole group, hand over the chalk.
- Review written work. Look beyond right and wrong answers to learn everything you can about what a student knows. Determine what seems solid and easy, as well as patterns in errors. If students are scattered, suggest ways they can organize their work; this is likely to lead to more efficient problem solving and clearer communication.
- Question. Hearing the right answer is not necessarily a cue to move on. Question students at this point too. Specific questions are included in the lesson facilitation.

## Unit Sequences and Connections

The sequence in which the *EMPower* units can be used effectively with your class will depend on the backgrounds and interests of students. The units are not numbered so teachers can order them according to their class needs; however, the authors suggest specific unit arrangements that will support students' progression through certain concepts.

The authors do not recommend sequencing the units according to the traditional basic math model that begins with whole numbers, follows with fractions, decimals, and percents; data and graphs; algebra; and then geometry. Instead, the authors suggest you integrate the five units that focus on numbers with the units on geometry, data, and algebra. The authors found this integration of topics helped to motivate the adult students in their pilot classes.

Although the units were not specifically designed to build on one another, there are clear connections between some of the units in the series. *Over, Around, and Within: Geometry and Measurement* provides a nice introduction to the program because it focuses on small whole numbers. *Everyday Number Sense: Mental Math and Visual Models* could follow to further develop whole number mental math skills and visual models. *Using Benchmarks: Fractions and Operations* provides the necessary groundwork with fractions, decimals, and percents to describe approximate relationships between data sets in *Many Points Make a Point: Data and Graphs.* And *Split It Up: More Fractions, Decimals, and Percents* continues to expand students' repertoire of familiar fractions, decimals, and percents. *Seeking Patterns, Building Rules: Algebraic Thinking* builds upon the tools and relationships used in *Keeping Things in Proportion: Reasoning with Ratios.*

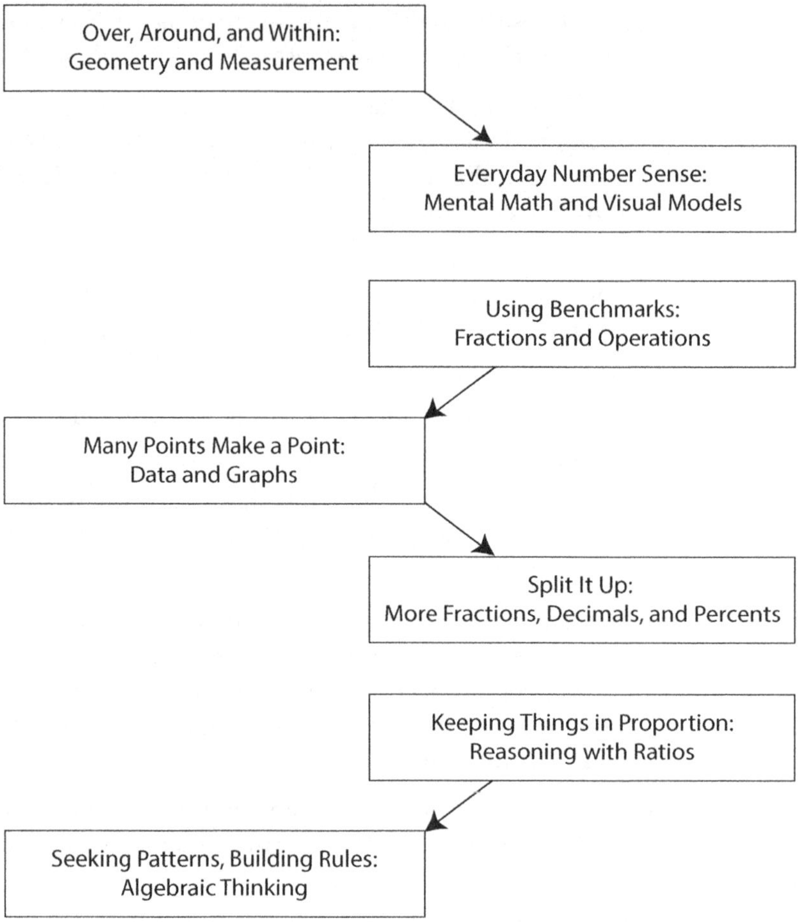


## Unit Descriptions

*Over, Around, and Within: Geometry and Measurement*
Students explore the features and measures of basic shapes. Perimeter and area of two-dimensional shapes and volume of rectangular solids provide the focus.

*Everyday Number Sense: Mental Math and Visual Models*
Students solve problems and compute with whole numbers using mental math strategies with benchmarks of 1, 10, 100, and 1,000. Number lines, arrays, and diagrams support their conceptual understanding of number relationships and the four operations.

*Using Benchmarks: Fractions, Decimals, and Percents*
Students use the fractions $\frac{1}{2}$, $\frac{1}{4}$, $\frac{3}{4}$, and $\frac{1}{10}$; the decimals 0.1, 0.5, 0.25, and 0.75; and the percents 50%, 25%, 75%, 100%, and the multiples of 10% as benchmarks to describe and compare all part-whole relationships.

*Many Points Make a Point: Data and Graphs*
Students collect, organize, and represent data using frequency, bar, and circle graphs. They use line graphs to describe change over time. They use benchmark fractions and

the three measures of central tendency—mode, median, and mean—to describe sets of data.

*Split It Up: More Fractions, Decimals, and Percents*
Building on their command of common benchmark fractions, students add thirds, eighths, hundredths and their decimal and percent equivalents to their repertoire of part-whole relationships.

*Keeping Things in Proportion: Reasoning with Ratios*
Students use various tools—objects, diagrams, tables, graphs, and equations—to understand proportional and nonproportional relationships.

*Seeking Patterns, Building Rules: Algebraic Thinking*
Students use a variety of representational tools—diagrams, words, tables, graphs, and equations—to understand linear patterns and functions. They connect the rate of change with the slope of a line and compare linear with nonlinear relationships. They also gain facility with and comprehension of basic algebraic notation.

## Frequently Asked Questions

**Q:** *I have classes that are widely multilevel. Can this work?*

**A:** Many teachers see a wide range of levels within the group as an obstacle. Turn the range of levels to your advantage. Focus on students' representations (words, graphs, equations, sketches). This gives everyone the chance to see that answers emerge in several ways. Slowing down deepens understanding and avoids facile responses. Having calculators available can even the playing field. Implement the suggestions in *Making the Lesson Easier* and *Making the Lesson Harder* of each lesson facilitation in the *Lesson Commentary* sections.

**Q:** *How do I deal with erratic attendance patterns?*

**A:** Uneven attendance can be disruptive. Students who miss class may feel disoriented; however, the lessons spiral back to the most important concepts. When the series circles back, students will have a chance to revisit concepts and get a toehold.

**Q:** *What do I do if I run out of time, and there is no way to finish a lesson?*

**A:** Each activity is important, but reviewing it is equally important. It is better to cut the activity short so there is time to talk with students about what they noticed. Maximize the time by selecting a student or group whose work you feel will add to the class's understanding to report their findings. Be conscious of when you are letting an activity go on too long because the energy is high. Fun is good, but be sure important learning is happening. If you like to give time in class to reviewing homework, and you want to hear from everyone in discussions, you will run out of time. Schedule a catch-up session every three or four lessons.

**Q:** *How do I respond to comments such as "Can't we go back to the old way?"*

**A:** Change is unsettling, especially for students who are accustomed to math classes where their job is to work silently on a worksheet solving problems by following a straightforward example. Be clear about the reasons why you have chosen to de-emphasize some of the traditional ways of teaching in favor of this approach. Ultimately, you may need to agree to some changes to accommodate students' input. Meanwhile, stick with the program. Reiterate for students what they have accomplished. When there is an "Aha!" moment, point it out.

**Q:** *My own math background is not strong. Will I be able to teach this series?*

**A:** Yes! Most teachers tend to teach the way they were taught. Adopting a different stance requires support, and the more types of support, the better. This series offers support in a few ways. The teacher books for each unit list open-ended questions designed to keep the math on track. In the *Lesson Commentary* sections, *Math Background* helps teachers deepen their understanding of a concept. In addition, the *Lesson in Action* sections provide examples of student work with comments that illuminate the underlying mathematics.

The best support often comes from a colleague. If no one at your site is currently teaching *EMPower*, join the Adult Numeracy Network, http://adultnumeracynetwork.org, and attend your regional NCTM conference. Look for others who are integrating the NCTM Principles and Standards through the use of a curriculum such as *Investigations in Number, Data, and Space* (TERC, 1997); *Connected Mathematics* (Lappan et al., 1998); or *Interactive Mathematics Program* (Alper et al., 1997).

# Unit Introduction
# *Many Points Make a Point: Data and Graphs*

## Major Themes

### The Importance of Data and Graphs

Even the most mundane decisions are based on *data*—information gathered through the senses as well as through systematic efforts to assemble relevant facts. To put data to good use, individuals need organizational, interpretive, and analytical skills. In this unit, students build these skills to communicate, influence, and make sound decisions and predictions based on data represented in tables and graphs.

### Interpreting Data and Graphs

As parents, workers, community members, and lifelong learners, adults need to be able to decode information that comes in bar, line, and circle graphs. Students become accustomed to asking themselves and answering questions such as

- What story can you tell from organizing and presenting data in graphs? What are the relationships among data points?
- What story do the data tell about change over time? Where are the high points, the low points, the greatest change, and no change?
- What impact does scale have on the reader's impression of the data and their understanding of the graph?

### Generating Data and Graphs

Interpreting a graph becomes easier if students first have hands-on experience constructing one. Construction requires close attention to the elements of a graph and a complete understanding of the process, from gathering data to a final representation. Categorizing information is a key step in planning a graph. It involves thinking flexibly about the characteristics of the data set. It also involves thinking about the purpose of an audience for the data.

The skill of generating graphs is useful beyond the insight offered through hands-on experience. Once students are able to make bar and circle graphs, they can discuss which format is more effective for making a point given a particular set of data. Creating a graph is another way to illustrate an idea. The ability to select appropriate information, organize the data, and construct graphs to communicate trends or influence decision-making is a valuable skill that has the potential to give a person confidence and authority.

### Decision-Making Based on Data and Graphs

This unit offers practice in evaluating data and graphs for the purpose of decision-making. Whenever possible the lessons include graphs based on real data, for example, climate changes, a credit card client's spending summary, kilowatt-hours, or health- and immigration-related graphs.

## Unit Goals

By the end of this unit, students should be able to

- Collect, organize, and represent data.
- Make accurate statements about data using percents and fractions.
- Create and interpret frequency graphs, bar graphs, circle graphs, and line graphs.
- Make predictions, decisions, and recommendations based on information presented in bar, line, and circle graphs.
- Use scale to change the story a graph tells. Interpret graphs with different scales.
- Use mean, median, and mode to describe a numerical data set.

## The Flow of the Unit

The *Teacher Book* provides materials for initial, interim, and final assessments (see *Assessments*, pp. xxiii–xxv, for more detail). During the first four lessons, the emphasis is on categorical data, that is, information that can be grouped or classified, such as types of foods or country names. In *Lessons 5–10*, the emphasis shifts to line graphs describing changes over time.

Each lesson has one or more activities or investigations. Allowing time for opening and summary discussions (including time for students to write reflections) and assuming a thoughtful pace, most lessons will exceed an hour. Students' reading and writing levels will have an affect on how long each lesson takes.

In *Lessons 1* and *2*, students

- Experience the process of collecting, organizing, and describing data.
- Use a frequency graph as a quick way to tally data.
- Generate statements using common benchmark fractions and percents to describe data pictured in frequency graphs.
- Regroup data and discuss the implications of the change.
- Speculate about a possible audience or purpose for the data.

In *Lessons 3* and *4*, students

- Transform frequency graphs into bar and circle graphs.
- Comment on the similarities and differences among formats.
- Sketch circle graphs based on division of a circle and estimates of percents.
- Use benchmark fractions and percents to describe data pictured in circle graphs.

*Lessons 5–10* turn to numerical data and representations of change over time. Students' skills quickly become more sophisticated as they handle graphs with two $y$-axes (two types of information) and compare graphs with different scales. Students will learn to

- Draw lines to match statements describing change over time.
- Precisely describe upward and downward trends and periods of stability.
- Match graphs and descriptions of climate data for five mystery cities.
- Use tables.
- Plot points.
- Develop strategies for finding the median and mean.
- Use mean and median to describe a data set.

When they reach *Closing the Unit: Stock Picks*, students are equipped to analyze five-year data on stock prices, compare and contrast the stock prices for two companies, and recommend one of the companies as the better investment.

### A Note on Number

Students need some facility with numbers when working with data.When they study a graph, they will not only look at its shape, but also notice relationships: "It looks like half of the people in that group are 35 years old or older."You want students to feel comfortable phrasing numerical descriptions of their data sets using benchmark fractions and percents (that is, the most common fractions, e.g., ½, ¼, ¾, and their percent equivalents). In this unit, students review and apply their working knowledge of these concepts.

If making numerical statements about the data proves difficult for your students, shift the focus. Return to the graph and ask "What do you see? When looking at a graph the shape is important; is it fairly even? Are there clusters? Are there gaps? Why might that be?" The conversations that arise about gaps, for instance, are often more telling than the data explicitly shown in the graph. Let numerical statements referencing "greater than" or "less than one half" suffice. Use this unit as a starting point to study fractions and percents.

## Assessments

*Many Points Make a Point* opens and closes with assessment. In both cases, the *Teacher Book* provides multiple ways to gauge what students know. In addition, the *Teacher Book* includes a interim assessment for students. There are also tools for ongoing assessment in each lesson. These components are described below.

### Opening the Unit

*Making a Mind Map:* Students record words and ideas they associate with data and graphs. By the end of the unit, their Mind Maps will likely include more terms, particularly technical ones.

*Initial Assessment:* Students check off "can do," "don't know how," or "not sure" in response to their ability to solve problems, assessing their familiarity with the material as well as their level of confidence. Checked against the *Initial Assessment Checklist*, the *Initial Assessment* will indicate which lessons will be new material and which lessons will be reviews.

*Sorting Graphs:* This activity is appropriate for students at all levels. Students examine and sort a collection of graphs, looking at their features. They place each graph in a category and make statements that describe the categories they chose. This is an informal way to assess what the class already knows about graphs and the vocabulary students use to describe them. This activity sets the tone for the unit by emphasizing pair work and communicating ideas.

*The Graph I Chose*: Students describe a graph, why it interests them, and what they learned from it.

### Interim Assessment

The *Interim Assessment* covers concepts in *Lessons 1–4*. Working in small groups, students select and analyze the data, choose a format to present their data to their classmates, and reach a decision or recommendation based on their data.

Students can also complete an independent project, wherein they collect and analyze data on foods eaten in a week. The project is assigned in *Lesson 2*. The data are collected and analyzed in *Lesson 3*.

### Closing the Unit

*Stock Picks:* Students evaluate graphs of two companies' stock prices and recommend the better investment. They use information in the graphs to frame a coherent, persuasive argument about the company they would choose, and possibly redesign the given graph (manipulating scale, for example) to make their cases stronger.

*Review Session:* Students review activities and practice sheets from past lessons and reflect upon what they have learned.

*Reflections:* Students organize a portfolio showing off their best work and commenting on the lessons that were most challenging.

*Final Assessment:* As they did in the *Initial Assessment*, students check off "can do," "don't know how," or "not sure," and then complete tasks and multiple questions, revealing their own assessment of their familiarity with the material, as well as their level of confidence. Checked against the *Final Assessment Checklist*, the Final *Assessment* provides information on which skills are solid and which will need further development. For their final Mind Map, students record words and ideas they associate with data and graphs. Their Mind Maps will likely include more terms, particularly technical ones, than were in their first Mind Map in *Opening the Unit.*

### Ongoing Assessment

Much of the work students do will take place in small groups or pairs. This work must be assessed to determine whether students are absorbing the main ideas and to diagnose difficulties, as well as provide more challenging work to students. The Looking Closely section of each lesson focuses teachers' attention on objectives and the corresponding observable behavior. Some teachers read this after they have taught the lesson and return to class prepared to assist students who struggled. Teachers may also revisit ideas that were not clearly understood.

As students gain skills, teachers will want to track their progress and communicate observations to them. Each lesson has a chart called *What to Look For in Lesson #,* with four dimensions for assessing students' progress:

*Concepts:* The student grasps big ideas and technical skills.

*Expressive Capacity:* The student has the ability to describe, reason aloud, and explain to others. This is easiest to observe in written work, a record with strong evidence of understanding and misconceptions or errors. However, a student's oral remarks are often a more accurate reflection of what he or she knows.

*Use of Tools:* The student has the ability to use calculators, grid paper, and $x$- and $y$-axes to communicate and enhance understanding.

*Background Knowledge:* The student has the ability to draw on and share past experiences using graphs to make decisions or predictions, or to influence or communicate information.

Many people talk about "drowning in data." In this unit, students will not drown. They will share knowledge gained in and out of the classroom to become proficient, discriminating users of data and graphs.

# Facilitating Opening the Unit: Many Points Make a Point

*What information does the graph communicate to you?*

## Synopsis

This session assesses students' familiarity and comfort with graphs. The activities establish a baseline by extending a variety of opportunities for students to interact with and comment on aspects of graphs, including topic, format, and purpose. The lesson introduces students to terms integral to discussing graphs in an organic way.

1. Students complete a Mind Map for data and graphs.
2. Students complete a self-assessment based on several problems.
3. Students work in pairs to categorize graphs and write statements describing their categories.
4. The whole class discusses the qualities all graphs share.

## Objectives

- Demonstrate prior knowledge of graphs
- Keep a record of vocabulary useful to the study of data and graphs
- Identify common features of graphs
- Form hypotheses about the purpose of the information, its intended audience, and its presentation

## Materials/Prep

- Colored markers
- Graphs
- Easel paper
- Scissors
- Staplers or tape

Gather graphs of all kinds. Include circle graphs, bar graphs, line graphs, graphs with percents and/or whole numbers, graphs with legends, and complicated and simple graphs on a variety of topics. Prepare one set of each graph for every pair of students. You may supplement your collection with the graphs on *Blackline Masters 1–3.*

Ask students ahead of time to bring in graphs. Prepare the *Initial Assessment* (p. 141) to display for the whole class as well as for each student to have one to write on.

## Opening Discussion

Provide students with a simple overview of the unit. Because decisions are based on data, it is important to critically evaluate information used to make decisions. Data and graphs can be overwhelming to someone unfamiliar with them. In the course of the next few weeks, students will begin to understand data and graphs by picturing, organizing, and making sense of information, and by considering its source. In addition, you and your students will construct different types of graphs and make statements, recommendations, decisions, and generalizations based on data from different sources.

## Activity 1: Making a Mind Map

The Mind Map is the first artifact of students' prior knowledge of data and graphs. By the end of *Opening the Unit*, you should have several artifacts:

- Mind Map (*Activity 1*)
- Students' *Initial Assessment* (*Activity 2*)
- *Sorting Graphs* statements (*Activity 3* in pairs or groups)
- Writing samples from *Practice: The Graph I Chose*

Start the Mind Map by writing "Data" and "Graphs" on the board. Ask:

**Where do you see graphs?**

**What do you know about them? List some of the characteristics of a graph.**

**For what purposes are graphs used?**

Write a few ideas. Model clustering them, linking related ideas with lines, and then ask students to record their own ideas on p. 2 of the *Student Book*. After a few minutes, ask students for examples from their work.

As terms related to data and graphs arise, start a class vocabulary list for graphs. Do not worry about formal, dictionary-style definitions; use what the students themselves say. You will want to add and refer to this word list, so write it on easel paper. Students can take notes in the *Vocabulary* section of their books, starting on p. 141.

## Activity 2: Initial Assessment

Because the *Initial Assessment* can look intimidating, it is recommended that you begin by explaining that this is a test where students first check off whether they feel they can do, don't know how, or are not sure whether they can do the problem. Tell your students that the results will help you decide what is most important to focus on in the unit on data and graphs.

You can complete the assessment as a group by displaying the problems one at a time and having the class read the problems together. Distribute a copy of the *Initial Assessment* to every student so that individuals can assess their confidence in their ability to solve the problems. Students can read silently or you can read to them. Keep the momentum going, reading the problem, and then moving on to the next one.

The *Initial Assessment* is a flexible tool. You can gain good information about students' skills simply by asking them to self-assess, indicating their confidence levels on the problems by marking them "can do," "don't know how," or "not sure" *without solving* them. Alternatively, students may solve the problems, starting with those that they feel they can solve and attempting those they are not certain about. Establish a firm time limit for this. Then collect the assessments and analyze them later. Students' work on the problems will provide a point of comparison for assessing their progress when they have reached the *Interim* and *Final Assessments.*

See the *Initial Assessment Checklist* (p. 146) for information on instructional decisions.

## Activity 3: Sorting Graphs

Pair students and distribute one packet of graphs per pair. Ask students to examine and sort the graphs into groups based on their similarities. Demonstrate by selecting a graph and naming the category you would place it in. For example, you might place a pie chart on diabetes in the category "circle graphs" because of its type or in a "health" category because of its content. Allow students 10 minutes to work on sorting the graphs.

Once pairs have made their decisions, allow up to 10 minutes for them to prepare a display and an explanation of what the graphs in each of their categories have in common. Provide a sentence starter or sample statement that students can use as a model, e.g., "Overall, the graphs in this category … "

Bring the whole class together to discuss their graph categories and reasoning.

## Summary Discussion

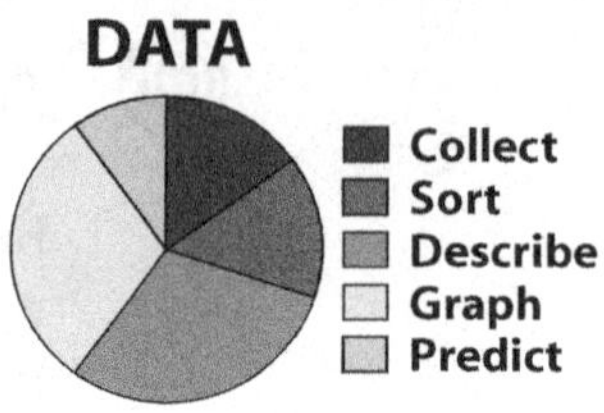


Facilitate a discussion about the sorting activity. Involve as many students as possible in explaining their reasoning for their categories.

Ask:

**What do your categories have in common with those that other pairs of students made?**

**How are your categories different from those of others?**

Focus on how the graphs display the information. Pick one or two graphs and ask pointed questions, such as

**What were some of the differences or commonalities in the content?**

**For what audience does the graph seem designed?**

**What do you think the purpose of the graph is?**

Summarize the characteristics students mention: "I heard you say..." Then ask:

**What do the graphs have in common overall?**

---

## Heads Up!

If your students all formed the same kinds of categories, you will need to be explicit about other features worth mentioning. For example, ask students to point to a graph where there is a dramatic difference pictured in some way.

---

Ask:

**What are some ways in which these graphs could be used?**

**How do you see numbers used in the graphs?**

Conclude by asking:

**What are some of the benefits of looking at graphs compared to reading an article?**

If there is time, orient students to their books. Students should answer the *Reflections* questions (*Student Book*, p. 144). Also have them read through the *Unit Goals* (*Student Book*, p. 6).

## Practice

*The Graph I Chose* (*Student Book,* p. 4)
This is the initial written assessment for the unit. It would be ideal to have a sample of each student's work. Therefore, allow class time to work on this if your students do not do homework reliably. See p. 12 for sample high-level responses. Ask students to

- Pick one graph that interests them,
- Save the graph with the practice page, and
- Write two paragraphs about the graph to answer the questions on the page.

*Data Collection* (*Student Book,* p. 5)
Ask students to collect data that will be central to *Lesson 1, Countries in Our Closets.*

---

## Heads Up!

Save the circle graphs and line graphs for *Lessons 4* and *5.*

---

## Looking Closely

Observe whether students are able to

### Demonstrate prior knowledge of graphs

Which students have experience with graphs or charts? What examples do they cite? Make note so you can draw on these students' knowledge during the lessons.

### Keep a record of vocabulary useful to the study of data and graphs

Introduce the *Student Book* format and point out that students have a place to keep track of vocabulary. Encourage them to make notes of new terms, using their own words.

Can students track horizontally and vertically at the same time—that is, are they looking at the graph and paying attention to information on both axes? Do they use words like "percent" when talking about circle graphs?

Pay special attention to whether students are able to paraphrase—e.g., "This graph is about… (they rephrase the title)." "This graph is telling me …" Are students able to use mathematical language to describe the data?

### Identify common features of graphs

Students need to look at more than one graph at a time. If a pair becomes fixated on one graph, encourage these students to get a sense of all the graphs. Brainstorm with students how they could group the graphs before they implement one strategy. Remind them that there is more than one right way to sort graphs. It is key that students articulate sound reasoning for their groupings.

### Form hypotheses about the purpose of the information, its intended audience, and its presentation

Do students seem to understand how graphs are used? Or have they tended to avoid graphs because they are unsure of how to interpret them? Welcome speculation about who might use the graphs and how they might be used to inform decisions.

| WHAT TO LOOK FOR IN *OPENING THE UNIT* | WHO STANDS OUT? (LIST STUDENTS' INITIALS) | | | NOTES FOR NEXT STEPS |
|---|---|---|---|---|
| | STRONG | ADEQUATE | NEEDS WORK | |
| Concept Development<br>• Distinguishes graphs from other representations (e.g., pictures, maps)<br>• Isolates information on graph<br>• Understands that time periods often play a role in graphs<br>• Recognizes fractions and percents on graphs<br>• Identifies examples of context where graphs are used | | | | |
| Expressive Capacity<br>• Uses vocabulary related to graphs (axis, scale, percent, etc.)<br>• Describes types of graphs (bar, circle, line)<br>• Summarizes content of graphs<br>• Self-assesses strengths and weaknesses ("I understand bar graphs.") | | | | |
| Background Knowledge<br>• Reports comfort level with different types of graphs<br>• Reports experience with graphs<br>• Relates content of graphs to prior knowledge | | | | |

## Rationale

This is an assessment lesson—an opportunity to get an idea of what your students know about data and graphs through their comments, explanations, and interpretations. This lesson invites students to think about graphs and to remember what they have learned about them in the past. Sorting graphs is a welcoming introduction to the subject for students with varied backgrounds. Do not dwell on *Activity 2: Initial Assessment*. Students should read the problems, self-assess, and move on to *Activity 3*. Problems similar to those on the assessment will be addressed throughout the unit. Teaching to the problems at this point will undermine the sequential development of skills and concepts.

## Math Background

The four types of graphs covered in this unit are bar, circle, line, and frequency graphs. Although they are sometimes used interchangeably, at times one graph is more useful than another. There is often more than one correct way to display a set of data. However, each type of graph has particular strengths, and sometimes one type of graph is useless for a particular purpose. For instance, it is impossible to use a circle graph to show how your body temperature fluctuates over time.

Circle graphs, also known as pie graphs, are excellent tools for making comparisons. Each slice, or category, can be visually compared to the whole circle representing the total. Sometimes the slice and the whole circle are thought of in absolute terms (e.g., 4 left-handed people compared with the total of 16 people); sometimes the slices and the whole are described in relative amounts (e.g., 25% compared to 100%).

Bar graphs can be used for the same purpose as circle graphs: to make comparisons among categories. Although the "whole" may not be as visually obvious as it is in a circle, the individual bar heights make it easy for readers to see the differences between categories. Bar graphs can display relative amounts (percents) or absolute numbers. When the amounts are absolute numbers, it is usually fairly easy to determine the total number of data items in each category because of the scale on the $y$-axis. Bar graphs can also show how amounts have changed over time. The format lends itself to tracking trends—e.g., the number of different car types that sold in each of the last three months.

Line graphs are ideal for showing change. Although they have other uses, in this unit, line graphs will be used to show trends over time. The $x$-axis is used for time in the graphs developed in these lessons—e.g., changes in attendance, stock price fluctuation, or temperature changes. The data points are interpreted using the scale on the horizontal and vertical axes. When the data are continuous, such as weight gained and lost, the line truly represents the data. When the data are discrete, such as prices, lines can connect the data points so that a trend is visualized, but in fact, prices jump. For example, it is not necessarily true that when gas prices climb, consumers can buy a gallon of gas for $3.29 and then $3.30, $3.31, $3.32, etc. It is more likely that one day the price is $3.29 and the next it is $3.35. The line shows the trend, but it is not a true reflection of the cost over time.

Frequency graphs are a quick way to tally and organize data, revealing features of the data such as gaps or clumps. One advantage is that they do not require time-consuming attention to color, design, labels, and intervals.

Students in pilot classes have asked, "What is the difference between a graph and a map?" You might want to be prepared to respond. A map refers to some type of representation of the earth's surface—e.g., a town's roads, boundaries between countries, or topographical information. A graph is a visual representation of information—data—in an organized form that allows for quick analysis. Graphs and maps both have coordinates and axes. But a map is ultimately a picture of a place rather than a representation of discrete data or information.

## Facilitation

Read aloud and explain the *Unit Goals* today (*Student Book*, p. 6), if there is time. If not, return to them later. The goals might have more meaning at a later time when students have more familiarity with the concepts.

### Making the Lesson Easier

Some students might have trouble understanding what is being asked of them. Instead of requiring them to choose their own groupings, you might ask them to sort their data into "Interesting/Not interesting." Display an example graph sort with two columns and two headings (e.g., "Interesting/Not interesting" or "Appealing/Not appealing") to help get them started.

To make writing statements easier, post sentence stems such as

"I like this graph because ..."

"I did not like this graph because ..."

# OPENING THE UNIT IN ACTION

*Notice the teacher's open-ended questions and her efforts to involve many students. She seeks to knit together the collective knowledge in the sheer volume of information from the students in this very multi-level class.*

Sylvia, the teacher, asked: "What is a graph?"

**Celine**: "Increase or decrease or change."

**Sylvia**: "Why do we have them? Where do you see them?"

**Melvin**: "Has a purpose."

**Annette**: "Magazines, news."

**Melvin** (pointing to map): "Is that a graph?"

**Sylvia**: "No. Anyone want to show what a graph is?"

Celine is at the board drawing two vertical lines. She labels one "1980" and the other "2000" and makes a horizontal on each.

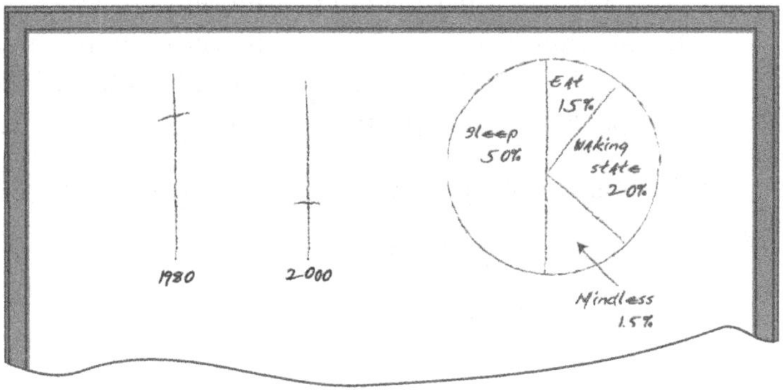


**Melvin**: "Up and down; the economy." (Holding his fingers up to show a bar graph), "Some companies have a monthly report."

Annette says she is tired, but she goes to the board to do a circle graph. Sylvia suggests Annette do a chart on sleep. Annette does and labels each section of the circle with a percent.

**Hanna**: "Every time I go check up for Aliya, she [the doctor] shows me a graph of height and weight." (Hanna goes to the board and draws two axes. Her *x*-axis has a very carefully drawn scale with dark lines periodically and about 10 smaller lines between each of these. She also writes "wt" on one axis and "ht" on the other axis.)

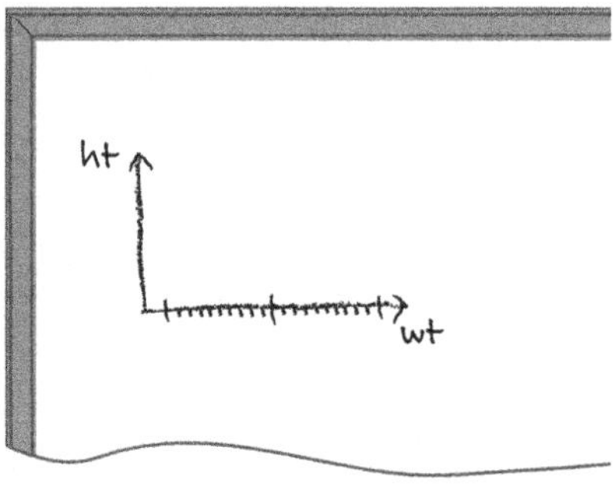


*(continued on next page)*

*(continued from previous page)*

**Hanna**: "She [the doctor] showed me a graph, but she did not make a line."

**Sylvia**: "Remember, Melvin, you said a graph has to have a purpose? What is the purpose? Why did she do height and weight together instead of weight on one graph and height on the other?"

**Annette**: "They are related."

**Melvin**: "Balance."

**Sylvia**: "Relationships. So the height and weight are related so they are on the same graph."

*Sylvia Lotspeich Greene*
*Community Learning Center, Cambridge, MA*

## Assessing Written Work: The Graph I Chose

The following are sample responses to *Practice: The Graph I Chose*. If most of your students' responses are as good or better than these, ask them to complete the problems on the *Initial Assessment*. Using the checklist, determine which lessons will be most relevant to them.

The criteria used to select these responses as being high level were concept development, expressive capacity, and background knowledge. You will see these same categories in a chart in every lesson. They are one way to record evidence of your students' progress.

### Concept Development

- Picks out important numerical information and interprets its meaning correctly
- Compares categories within the graph
- Uses the information to make a statement
- Demonstrates an understanding of graph construction

### Expressive Capacity

- Able to summarize the content of a graph
- Able to clearly paraphrase the content
- Uses precise vocabulary to describe changes or patterns in the data

### Background Knowledge

- Connects information in graphs to own life
- Connects to own world view

## Sample High-Level Responses

These responses were taken verbatim from student work in several EMPower classes.

### *The Graph I Chose*: Fun and Games Online

*What do you know or what did you learn about the data represented in the graph?*

**Student 1**: "I learned that this year while going online sales are expected to reach $790 million, which pertain to toy and video, as people purchase their merchandise online. On the priority list [the] majority [of] folks purchase books, toys/games, apparel and footwear, music CDs fourth, and wine and gift[s] ending last …

*Evaluate the graph.*

"I think this survey is excellent, as folks could kick back saving time and energy not having to make several trips seeking gifts. This survey [in] my viewpoint would convince people to shop smart save cash."

*What do you know or what did you learn about the data represented in the graph?*

**Student 2**: "I learned that many more people are willing to shop online than I had imagined. I also learned that books and toys were the favorite online gifts to buy. The last thing I learned is that there are price differences from site to site, but I would wonder about reliability too."

*Evaluate the graph.*

"This graph was cluttered and was really enough material for three graphs."

### *The Graph I Chose*: Close Race

*What do you know or what did you learn about the data represented in the graph?*

"It shows how close the race really was. It shows the incline and decline of both candidates for both months. All three have almost the same exact percentage.

It also shows why it took so long to determine who our president was going to be."

*Evaluate the graph.*

"I think the presidential race was too close therefore I demand a recount. And Al Gore would be president. It was anyone's race all the way to the end."

### *The Graph I Chose*: Learning to Use Computers

*What do you know or what did you learn about the data represented in the graph?*

"I learned that most people learning how to use computers are self-taught, 26% people learned at school, 23% people learned through work and 6% people learned from family and friend."

*Evaluate the graph.*

"I [am] interested [in] it because I like [to] use computer[s] and I am one of the 41% people who are self-taught."

### *The Graph I Chose*: Smoking-Related Deaths

*What do you know or what do you learn about the data represented in the graph?*

"That there is a higher risk of dying from smoking than many other things. For example, there were about 247,700 deaths from alcohol, car accidents, AIDS, suicide, homicide, fires, cocaine/crack, and heroin. Smoking or using tobacco products had almost double the death[s] in one year, which was 435,000."

**Smoking–Related Deaths**

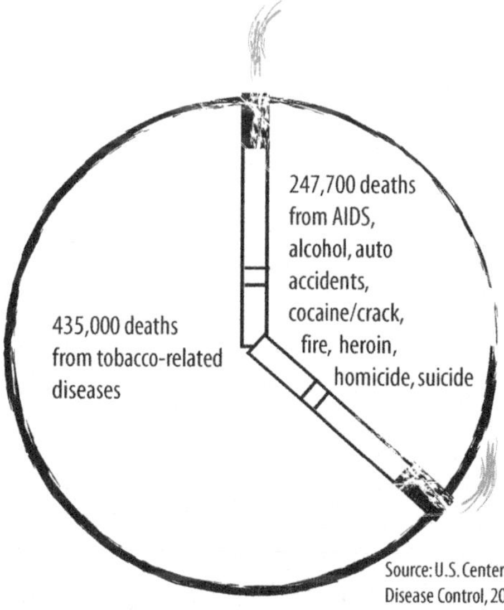


Source: U.S. Centers for Disease Control, 2000

More Americans die from smoking every year than drugs, AIDS, car accidents, alcohol, suicide, homicide, and fire combined.

*Evaluate the graph.*

"I liked the graph because of the information they listed but they could have listed more. For example, what year did they receive this information and compared to a recent year. It does have a good effect on me, however, all drugs have different effects on people and how and why many people do them what the graph should be based on. I also see the point of the graph is for people to stop smoking."

# FACILITATING LESSON 1

# Countries in Our Closets

***Where were your clothes made?***

## Synopsis

This is the first of a set of lessons on categorical data in which students collect, examine, sort, and organize their own data, and then compare their data to another data set.

1. Students bring to the class data on the different geographic locations where their clothes were made.
2. The class combines its data into one frequency graph, and students make several statements about it.
3. Working in pairs, students compare their class graph to data from another class.
4. Pairs of students each create a new frequency graph, organizing the class data by continent. Students notice which continent is most prevalent.
5. Discussion centers on comparing country and continent graphs, closing with the question How does the story change when the graph's categories change?

## Objectives

- Use a frequency graph to organize data
- Identify the story that the data tell
- Compare data from various samples
- Change the categories and articulate the change in the story

## Materials/Prep

- Grid paper
- Paper strips for posting statements
- Sticky notes
- Rulers

As an aid for determining in which continent the countries are located, have access to a world atlas, if available. For a listing of countries in Asia, see *Resource: Countries in Asia, Blackline Master 4.*

Make sure you have assigned *Countries in Our Closets* (*Student Book*, p. 7). Collect and bring in about 20 items of clothing in case your students have not collected their own data. (See *Facilitation*, p. 25, for how to handle this.)

Prepare to display the graph "Countries in Our Closets, Springfield, MA" (*Student Book*, p. 10) during the opening of Activity 2.

## Opening Discussion

Ask students:

**Are clothes made in this city? In this country? Why or why not?**

**Why would anyone care where his or her clothes were made?**

Clothing is close to us, physically worn next to our skin, so we should want to know where it comes from, but sometimes we take it for granted. We need to know something about where our clothes come from because of human rights, political, economic, and environmental issues. We can consider, for example, the age, safety, and wages of people manufacturing clothes in countries with fewer worker protections; the economic implications of outsourcing; and the cost of fast fashion to the resources of our planet.

---

### Heads Up!

If students do not bring in data, see *Facilitation*, p. 25.

---

**You have looked in your closets, checked the tags on some of your clothes, and collected information. What did you learn?**

**Which country do most of the clothes you checked come from?**

Post the country names in a place where you can easily refer to them later.

Introduce the terms **data set** and **mode**. Each student's list is a data set, and each data set could have a mode—the country name that occurs most frequently.

*Vocabulary note*: If students are searching for words to express their thoughts, offer terms and concise definitions. Students can keep track of these in their student books. For example, **data** are the information collected. The data that are collected make up the **sample**. The whole group from which information could be collected is the **population**. See *Lesson 1 Commentary*, p. 24, for more information.

## Activity 1: Organizing the Data

Explain the motivation for organizing data: It is easy to identify the mode from a small sample, but when working with larger samples of data, it helps to have a strategy. In addition, a graph will illustrate the data, enabling everyone to see the number of articles of clothing from each country.

Distribute sticky notes to students. Ask them to write on a separate sticky note the name of the country each article of clothing came from. On the board, draw a horizontal line, long enough to fit many different country names. Ask students to place their sticky notes above the line, organizing the data by country. Volunteers should write the names for countries along the bottom of the line. This graph is called a **frequency graph** because it shows how often (frequently) each country occurs.

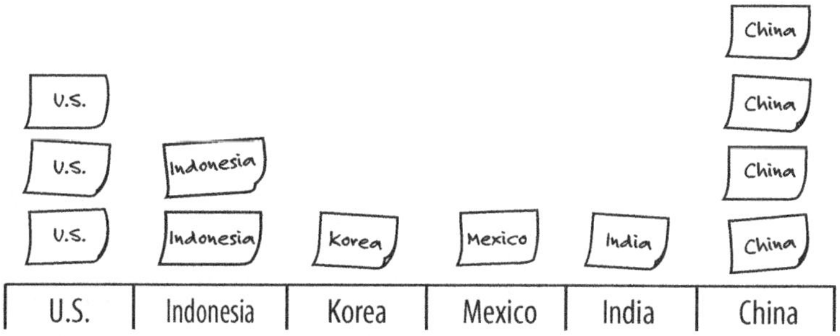


For distance learning, consider using a platform like Jamboard to collect and organize data for Activities 1-3

---

### Heads Up!

Leave the whole-class frequency graph posted for use in the following activity. If you have a copier available, make and distribute to each student a copy of the class frequency graph done on graph paper.

If you do not have a copier, distribute graph paper for students to copy the whole-class data for themselves—be aware that this could be a time-consuming task.

---

When students finish, ask:

**Which country did most of our clothes come from?**

**How does the mode for the whole-class data compare to the mode for your own data?**

Ask students to create a few paired statements about the class data. Model this with a statement such as

**One article of clothing was made in each of the following countries: China, Mexico, and Korea.**

**There are twice as many clothes from China as from Mexico.**

Referring to the whole-class frequency graph, you might say:

**All these sticky notes could fly off, and we would lose our data. One way to avoid that problem is to use x's in place of each sticky note like this (model it):**

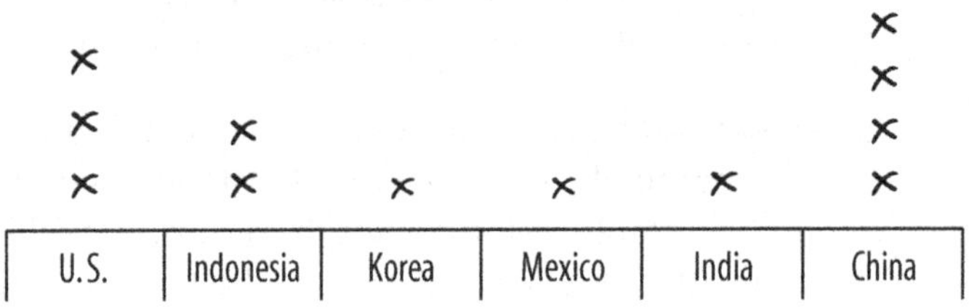


Note the components of a frequency graph:

- Line
- x's
- Labels

## Activity 2: Statements about Data

Students now compare their class data with data from another class. Post the data and ask students to comment on similarities and differences between the country data from their class and from a pre-GED class in Springfield, Massachusetts.

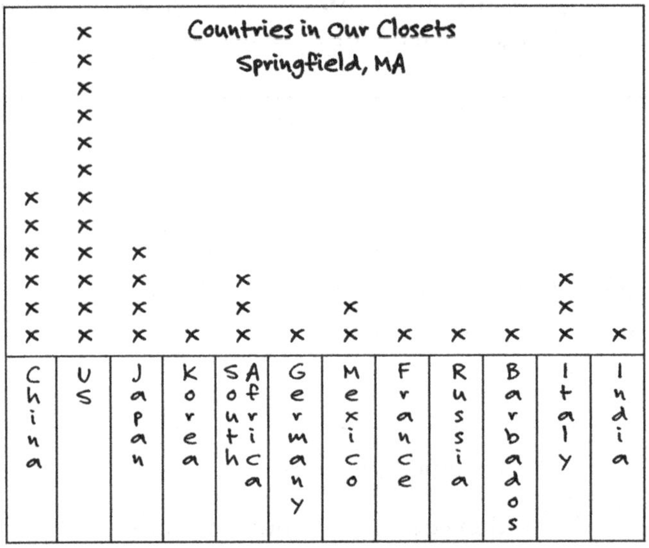


Start by describing the Springfield data:

**In the Springfield data, four times as many clothes come from the United States as from South Africa. Explain that statement.**

As students explain, make sure "3 x 4 equals 12" is made explicit.

**What other statements can be made about the Springfield data?**

Prompt students to provide numerical examples (e.g., twice as many clothes came from China as from Italy; half as many clothes came from China as from the United States; nine more articles of clothing came from the United States than from Italy). Refer students to *Activity 2: Statements about Data* (*Student Book*, p. 10). Ask them to work in pairs to answer these questions:

**We have data from two sets of students—one in Springfield and our own. How are they similar? Different?**

**Do our two classes have the same mode? Why do you think that might be?**

**What if we asked students from another state to make a data set? How do you think their data would compare to these two data sets? Why?**

**What if we included people other than adults in our program? How do you think their data would compare to these two sets?**

When researchers collect data, they ask themselves these kinds of questions.

## Heads Up!

The important thing to note is that the statements made are valid *only* for the group from which the data are collected. We can hypothesize that other groups might have similar or different data, but we do not know for sure unless we collect data. This is meant to open a conversation about the concept of sample. Draw from the *Math Background*, p. 24, if your students are interested. *Lesson 2* addresses the issue of sample in more depth.

## Activity 3: Changing the Categories

Next, the data will be grouped in another way—by continent. Students will work with the class data and might need additional reference materials. Refer students to *Activity 3: Changing the Categories* (*Student Book*, p. 11). Ask pairs of students to make a frequency graph showing the continents in their closets.

**When you make a frequency graph, remember the three parts: the line, the x's, and the labels.**

Ask the student pairs to post their graphs and to make one numerical statement describing the data in their graph. For example, "Less than half of the clothes came from Europe"; or "A quarter of our clothes came from South America."

**What do you notice about this new way of sorting our clothes?**

Be sure to highlight the following:

- It is often easier to remember and manage fewer categories.
- All the clothes from one continent could have been produced by many countries or by just one. The continent label hides this information.
- The "part of the world" many of our clothes came from becomes clear.
- The continent labels might change the relationships among categories. For example, the frequency graph by country could show that more than half the clothes were made in China; whereas the graph by continent could show that more than three-quarters of the clothes were made in Asia.

Open a discussion about the sample:

**What do you imagine you would find if a different group collected and shared their data?**

**Would it be appropriate to make statements based on these data that would apply to all adults' clothes? Why?**

---

## Heads Up!

Generalizations from data are affected by how representative the sample is and by its size. You do not need to go into details about this now, but if students wonder about whether they *can* generalize, give an example such as "If all our data were collected in (*name a wealthy area*), the countries might not be the same as if we collected data from (*name a non-wealthy area*)"; or, "If we only looked at clothes bought at (*name a high-end clothing or a discount-fashion store*), we might be missing some countries."

---

## Summary Discussion

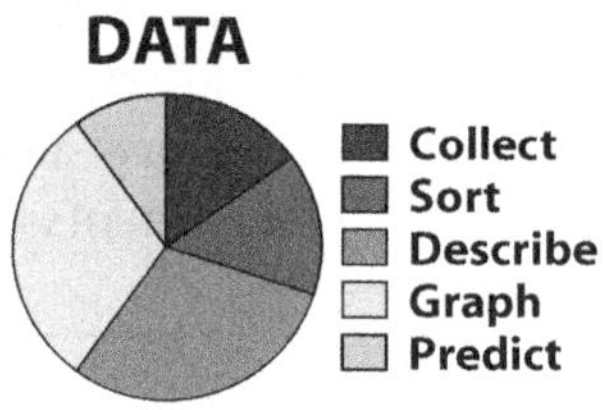


Encourage students to share—orally or in writing—what they have learned about collecting, sorting, graphing, and grouping and regrouping data using a frequency graph.

Say:

**You are in charge of making statements. The data are just pieces of information until you interpret them. The data stay static (that is, the data do not change)—the way they are represented does not. You can change the categories and/or combine them; this might yield a new story.**

Circle back to the larger conversation about where the clothes were made. The data will likely show that most of our clothes were made far from home. Engage students in a conversation about why this is. You might jump-start the discussion by saying:

**Although we have not talked about the cost of shipping or the cost of each item, companies seem to want to buy clothes made in other countries. Why might that be?**

You might also pose the question, "What does 'Made in the USA' really mean?"

This is a good time to remind students about the three components of a frequency graph. Go over the steps they took to represent the data, i.e., collecting, sorting, organizing, labeling, and making statements and generalizations about the data. Direct students to *Reflections* (*Student Book*, p. 144), where they can record their answers.

## Practice

*Clothes by Continent in Springfield, MA,* p. 12
For practice sorting data by continent.

*Reporting Data 1 and 2,* pp. 13–14
For practice seeing what effect collapsing data has on the story the graph tells.

*Categorically Speaking,* p. 15
For practice analyzing how things in our lives are organized in categories.

## Extension

*Taking Inventory,* p. 16
For practice collecting new data, then sorting, categorizing, and reorganizing them.

## Test Practice

*Test Practice,* p. 17

## Looking Closely

Observe whether students are able to

### Use a frequency graph to organize data

How do students manage the logistics of constructing a frequency graph? Make sure they use the grid lines as a guide so that that the x's are uniform. This way, when they compare countries, the comparison will be precise. Help students make the connection between one article of clothing and one x. Ask them to double-check their graphs to make sure they included every data point, not adding or omitting any. The category labels are also important. Assist students with spelling, if necessary.

### Identify the story that the data tell

Do students make statements based on the data displayed on the graph? Some students might feel the job is done when the graph is complete. Not so. If students are hesitant, pose questions about what the data tell: "Is there one country that stands out in the graph? How so? Why might that be?" Accept either statements that quantify the relative value ("Asia has three times as many") or the total value ("India has nine more than …").

### Compare data from various samples

Do students notice the big differences or similarities between two data sets? If they say, for example, "This data set includes Japan and the other one does not," that is a good start. Probe for more by asking, "Do they both have the same mode? How do you know?" You are only looking for a big-picture comparison.

### Change the categories and articulate the change in the story

Do students recognize the differences between the frequency graphs and the accompanying stories for countries and continents?

While some of the graphs might look similar, the labels are different, as are the groupings. Ask students who have trouble articulating specific questions: "What are the categories (groups) in this graph? And in the other one? Are there continents that stand out? How are they related to the countries that stood out?"

| WHAT TO LOOK FOR IN *LESSON 1* | WHO STANDS OUT? (LIST STUDENTS' INITIALS) | | | NOTES FOR NEXT STEPS |
|---|---|---|---|---|
| | STRONG | ADEQUATE | NEEDS WORK | |
| Concept Development<br>• Records all data<br>• Organizes data<br>• Recognizes key features of frequency graph<br>• Compares data<br>• Changes categories and notes change in story | | | | |
| Expressive Capacity<br>• Uses precise language to describe the frequency graphs<br>• Contributes to the discussion about the way changing categories affects the story the data tell | | | | |
| Use of Tools<br>• Uses graph paper<br>• Uses atlas or other reference | | | | |
| Background Knowledge<br>• Draws on understanding of geography, relative size, and number sense for comparisons | | | | |

## Rationale

The lesson takes students through the first steps of displaying data: collecting and organizing the information. As you work through this lesson, you and your students will notice that each time you categorize and recategorize the data set, you tell a different story.

## Math Background

Data are information. This information may be numerical—e.g., salaries, test scores, heights, age, weight—as well as categorical—e.g., countries, foods we eat, types of product defects.

Most data are reported using statistics based on fractions and percents—one-third of pregnant mothers, half of native plants, etc. Commonly used, or benchmark, fractions and percents are often invoked to influence others to make decisions. When the headlines shout, "One-half of all pregnant women …," readers tend to form opinions based on the data. What readers often neglect to consider is the size of the sample, where the data were collected, and how the data were organized.

The media generally report data that are based on a sample. It is rare for any group to have the time or resources to ask every person in the population to contribute information. The idea of sampling is to study a part in order to gain information about the whole. The sample—its size and characteristics—influences the data and the conclusions of the study.

A representative sample by definition includes representation across the population being surveyed. For example, a representative sample of the community would include individuals from across town, not just one neighborhood. Likewise, a representative sample across America would include individuals from many different states and regions, not just one or two. The size of the sample is important when inferences are made. For example, 30 samples could be sufficient for a population of several hundreds. The deciding factor for determining sample size is how confident you want to be about the inference.

It is not the concern of this unit to determine appropriate sample size, but it is important to note that statisticians use formulas derived from repeated surveys to decide sample size. Regardless, small samples are used in many studies. Factors such as availability of subjects or scarcity of time or financial resources might lead researchers to use a small sample and to base policy or claims upon the outcome.

Throughout this unit, it will be important to help students begin to think critically about data—whether the data are grouped into categories of information or into graphs. They will need to understand that a random sample will often yield different results from one that is not random. This topic resurfaces in *Lesson 2*.

## Context

Some students may know about *maquiladoras* in Mexican border towns, where women make clothes for very little money and with no benefits or environmental Occupational Safety and Health Administration (OSHA) workplace protections. CorpWatch (www.corpwatch.org) is one source for information on *maquiladoras*.

## Facilitation

If students do not bring in data, or if their sample is too small, skip the second part of the *Opening Discussion*. Have available a pile of 20 clothing articles with labels. First, ask students to predict where the clothes were made. Post the list of their guesses. Note that it will be hard for them to answer this question unless they organize the information on the labels. Then divide up the 20 articles of clothing. Have students write the name of the country for each piece of clothing on a sticky note, one country name per note. Ask: "Where are most of our clothes made?" Then continue with the activity.

### Making the Lesson Easier

Frequency graphs lend themselves to comparisons among categories. If students have little fluency in stating comparisons, you may choose to compare only size, using terms like "greater," "fewest," or "less than." For students who are encountering data formally for the first time, the notion that collapsing data yields different stories may be difficult. Treat this lightly in the activity, and revisit such questions after students have more experience categorizing and recategorizing data in the homework and in *Lesson 2*.

### Making the Lesson Harder

If your students understand benchmark fractions and percents such as one half or 25%, get them to look critically at the data, including the source and sample size. You might ask:

**If we asked another class what countries are in their closets, what do you think would happen to the categories? What if we asked the entire community?**

**How do you think your data would compare to data from another class of adult students in another community?**

If students struggle with the idea of sample, you might try this: Have them each write their favorite color on a sticky note. If you have a small class, ask them to write the color on two sticky notes. Place all of the notes in a container. Have someone randomly (eyes closed) choose a few notes from the container and place them across a line to form a frequency graph. Ask the students how they think this sample compares to the actual total number of colors on notes in the container. Students can create another frequency graph using all of the notes and compare the sample to the actual total.

# LESSON 1 IN ACTION

*Alice articulates the mathematical principle behind compressed data.*

I asked, "How did the change in categories affect what we noticed about the data?"

Alice answered, "Well, we keep losing information."

"How so?"

Patiently, Alice explained that when we started our work, every bit of data was visible. She added that we had lost details initially recorded. "At first, we knew every country in every person's closet and how many pieces of clothing came from that country. Then we combined the data, and we lost track of who had which countries. Then we did it by continent, and we lost track of all the countries."

Alice's realization quickly gained agreement from the rest of the class. After all, just the previous week a classmate had noted, "When you change the amount of data you look at, you find different things."

Sonia added her comment with increased conviction: "It is like politics. Politicians use a graph and tell you this is true, but you look at the graph, and it does not tell you everything."

*Tricia Donovan*
*Pioneer Valley Adult Education Center, Northampton, MA*

FACILITATING LESSON 2

# Most of Us Eat...

*What foods do you eat most often?*

## Synopsis

In *Lesson 2*, students again generate and organize data, but this time they choose their own categories. They use fractions and percents to describe the data, as they will again in *Lesson 4*.

1. Students think about the audience: For whom is data on food relevant?
2. Small groups of students organize their data and title their categories.
3. Back in the whole class, students decide on category titles.
4. The whole class describes the data using benchmark fractions and percents.
5. Pairs of students or the whole class create circle graphs with the food cards.
6. The whole class revisits the question of audience and the impact of categories and organization on the story the data tell.

## Objectives

- Organize data into consistent categories
- Organize data into categories to suit a purpose
- Make numerical statements about data

## Materials/Prep

- Notes (e.g., index cards, scrap paper, sticky notes, preferably different colors) or digital
- Sample chart to aid in making statements (see *Making the Lesson Easier,* p. 37)
- Strips of paper or large sticky notes, longer and narrower than the notes
- Tape, as needed

---

### Heads Up!

The lesson assumes students are familiar enough with benchmark fractions and percents to make statements using them. If this is not the case, see *Making the Lesson Easier,* p. 37.

---

## Opening Discussion

Begin by explaining that there are many studies and surveys that look at what Americans eat.

**Who would want to know this information?**

Advertisers, nutritionists, and health insurance companies might be interested. Invite students to contribute their own ideas.

**Who would be curious about what the people in this class eat?**

Pose this question and come back to it in the summary. The goal is to collect information about what this particular group eats. In *Activity 1,* each student will write on an piece of paper or sticky note the food they eat most frequently.

---

### Heads Up!

You may want to specify "food" as opposed to "drinks." Tea, coffee, and cola are derived from plants, but have no nutritional value. You may choose to have a quick discussion or simply announce a working definition of food so that everyone is in agreement.

---

## Activity 1: Frequently Eaten Foods

For this activity to work well, you need a data set of 16–20 pieces of information. If you have a large class, each student can volunteer one answer. If you have a small class, ask each student for two or three foods they eat most frequently.

---

### Heads Up!

Decide early on how many groups of students you will have. Each group needs a complete set of data, and you will need a set for *Activity 3*; therefore, each student will write down his or her food of choice as many times as there are groups of students plus one—each time on a different piece of paper or sticky note.

---

Instructions for students:

**Write the name of a food you frequently eat on a note. This is data. We will need a set of data for each group of students, so write a duplicate note for each group and also one for me.**

Collect one note from each student and say:

**Great, but I have so much data that I will never remember it all. I am going to see if I can organize it quickly.**

---

### Heads Up!

If you have plenty of class time—longer than an hour—ask students for their suggestions.

---

Model one way of organizing the data—e.g., raw food vs. cooked food, healthy food vs. unhealthy food, or ingredient vs. prepared food.

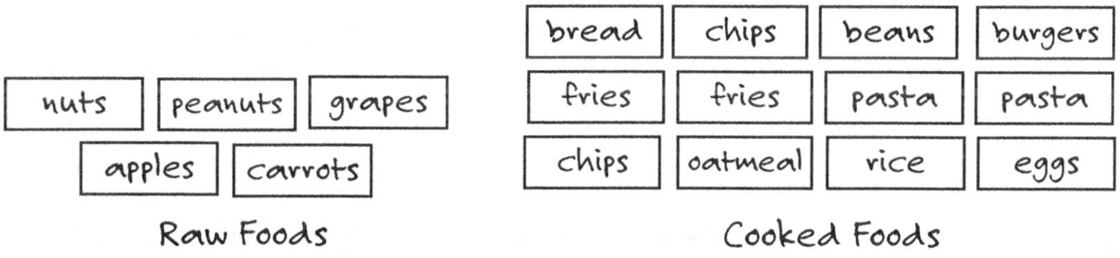


Keep track of the number of foods in each category. For example,

| Raw Foods | Cooked Foods |
|---|---|
| 5 | 12 |

Now ask students to work in small groups to create their own categories. They can make two, three, four, or even five categories with the data. Make sure every group gets a complete set of data.

### Organizing the Data

Each person takes a turn grouping the pieces of paper. Meanwhile, a group member writes the category names on a strip of paper. Another group member keeps track of the categories for each grouping, as well as the total in each category (e.g., raw food/5, cooked food/12).

After each student has taken a turn grouping the notes:

- Ask each group to choose *one* way of organizing the notes to share with the whole group.
- Ask each group to display the summaries of their other categories with the number in each category (e.g., raw food/5, cooked food/12).
- Encourage students to look at all the data discplays and ask questions or share comments.

There are four ways to capitalize on this activity: (1) by reflecting on the process of establishing categories; (2) by reviewing the groups' categories; (3) by making statements with fractions and percents; and (4) by considering the audience and sample.

## Activity 2: Consistent Categories

Ask students to think about the following:

**What made it difficult or easy to group items into categories?**

**Do you think there is a right or wrong way to organize the information? What is your reason for thinking so?**

Point to the strips that identify the categories, and read through them. Ask students to notice the variety of ways that the information can be organized.

Then look at some of the groups' displays. Point out that each organization gives a different picture of who we are and how we eat. Choose a group's display to discuss. Ask:

**What do you notice about how this group organized the data?**

Sometimes groups choose categories that are problematic because they overlap or are not consistent. Ask students about category titles that do not go together:

**What would be wrong with these categories:**

**Fruit, vegetables, meat, dairy, and tropical fruit?**

**Car, bike, bus, subway, highway, and carpool?**

Ask students to do *Activity 2, Consistent Categories* (*Student Book*, p. 21), and then revisit their displays and category names. Ask:

**What should you keep in mind when deciding whether your categories work?**

**What are some things to watch out for?**

Record and display students' ideas. Include on the list:

- The categories should fit most of or all of the data.
- The group should place the data in categories without disagreement.
- The data should fit one category only.
- New data, if any, could easily fit in one of the categories.

## Activity 3: Describing the Data

Looking at one of the groups' displays with just two categories, say:

**Try to use some fractions in making statements about the categories.**

**Take a look at these two categories and compare their sizes. Does one category include more than a half, less than a half, or about a half of all the data? Why?**

Refer to past lessons on fractions and percents, if relevant. Create some fractions together. For instance, in the example given earlier, $5/17$ of the most frequently eaten foods are raw, and $12/17$ are cooked. Establish the **benchmark fractions**. In other words, determine what "about half," "more than half," and "less than half" would be for the data set.

Ask students to complete statements such as

**The foods that we consume most often are ...**

**Over half the foods we eat are . . .**

Since categorical data are often reported with percents, and circle graphs quickly communicate percents of a total, design a circle graph with the extra set of cards. Arranged by category, all the cards should form the outline of a circle.

Review **benchmark percents** of 50%, 25%, and 75%. Draw a circle on the board, and have a student draw a line to show $1/2$, or 50%. Another student can show half of that—$1/4$, or 25%. On a second circle, ask a student to mark 25% and to label what is left. Ask:

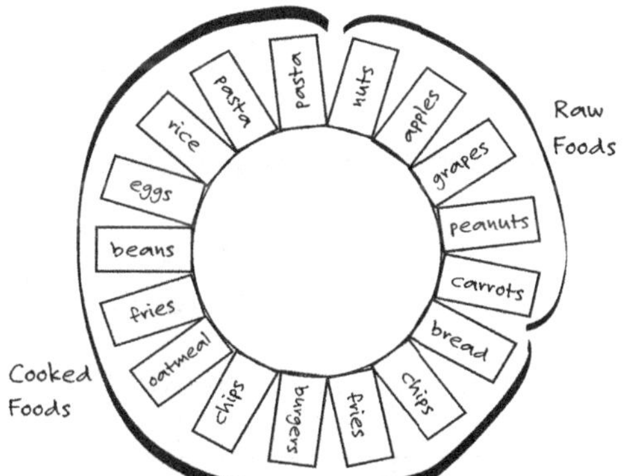


**What percent is the whole circle?**

**How many is the total in this case?**

Refer students to *Describing the Data* (*Student Book*, p. 23). Students describe the data either orally or in writing on this and other posters, using benchmark fractions or percents and statements such as "almost 25%," "a little less than 50%," and so forth.

## Activity 4: Who to Ask?

Ask students to make statements that highlight the differences between categories and to summarize the story of the displays. For example, say:

**In this organization of the data, it looks like most of us eat junk food. In this one, it looks like…**

Point to another group's display and ask your students:

**What can you say about our frequently eaten foods based on this display?**

Return to the brainstormed list from the *Opening Discussion* of who might care about what people eat. Ask:

**Who would find this information useful? Why?**

**How might ________ (e.g., nutritionists or health insurers) use this kind of data?**

**Which set of categories would be most useful for _______ (e.g., deciding what to stock in the school vending machine or what to bring to a class party)? Why?**

If students' categories are largely the same, pretend you are a client interested in some of the survey information the students have gathered. For example:

**I am stocking a new vending machine for the school. What data would I be interested in? What would you call the categories?**

**I am preparing a report on the nutrition of students in this school. What data would I be interested in? What categories of data would I want?**

**I want to do a presentation to students about avoiding colds by eating foods high in vitamin C. What data will help me decide whether students need this kind of talk?**

**Some people would be upset if a researcher or health insurer used this data to make a determination for a new policy nationwide. Why do you think that is?**

**What would you want to know before setting new policy?**

**Describe the sample, the group on whose habits you would want to base your research.**

Students work on *Who to Ask?* (*Student Book*, p. 23).

---

### Heads Up!

As with the data on countries producing clothing, the statements are valid only for the group from which the data are collected. Students may speculate about others' behavior based on what they themselves know, but the only way to know for sure would be to collect more data.

---

## Summary Discussion

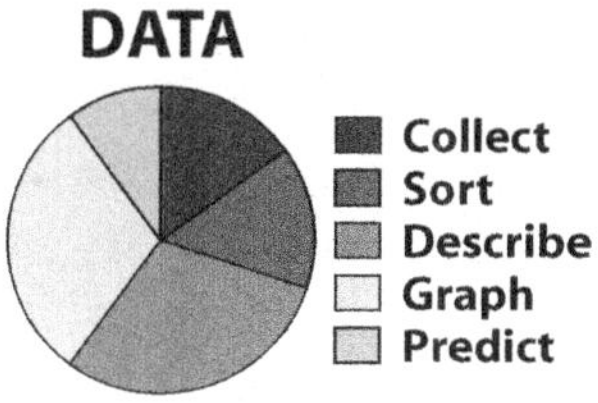


Finish the lesson by posing the following questions. (Students may not have ready answers at this point, but the questions will be a thread through the rest of the lessons.)

**What stories can be told from this set of data (or what questions can be asked and answered)?**

**How does organizing the information into different categories impact how the data may be interpreted?**

**Who would be curious about what the people in this class eat?**

Direct student to the *Reflections* questions (*Student Book*, p. 145).

## Practice

*Take a Sample,* p. 24
For practice choosing a representative sample.

*Thirsty* and *Ouch!,* pp. 26 and 28
For practice organizing information and making statements.

## Extension

*Friends and Family Drink Too,* p. 30

*Data Collection,* p. 31
Students begin recording what they eat for a project they will do in *Lesson 3*. For now, they just write down all the foods they eat throughout the day.

## Test Practice

*Test Practice,* p. 32

## Looking Closely

Observe whether students are able to

### Organize data into consistent categories

Do students group similar foods together? Make sure students are moving pieces of paper around and that they are writing down the category names. If students seem stuck, have them ask you specific questions.

If you see foods categorized incorrectly, ask a question such as "Do all group members agree with the categories?" Encourage students to think, question, and defend their reasoning. If students use overlapping categories such as "cereal" and "grain," present an example close to their own experience. Say, for instance, "It is confusing if I ask you to line up, and I say immigrants here and women over here.

Where do you go if you are both?" Have the students try to name categories so that no food belongs in two places. Note problematic categories and discuss them as a class during *Activity 2*.

### Organize data into categories to suit a purpose

How do students respond when asked how the data might be useful to others? If students struggle with the idea that the way information is organized can influence how it is interpreted, begin with one of the charts where most of the items are in one category. Ask them to think about what a nutritionist, for example, might say about this information.

### Make numerical statements about data

Key to interpreting graphs is the ability to detect mathematical relationships among the data. For example, when students see that one category is double the size of another or that two of the categories together make up half the total number of data points, they can more easily draw conclusions about the data. Begin this process with students, paying attention to how difficult it is for them. This is a multi-step task. If it is new for students, focus their attention on comparing quantities in different categories against the total.

| WHAT TO LOOK FOR IN *LESSON 2* | WHO STANDS OUT? (LIST STUDENTS' INITIALS) | | | NOTES FOR NEXT STEPS |
|---|---|---|---|---|
| | STRONG | ADEQUATE | NEEDS WORK | |
| Concept Development<br>• Organizes information<br>• Generalizes for titles<br>• Is able to use part/whole<br>• Has strategies for finding percents<br>• Makes reasonable numerical statements based on data | | | | |
| Expressive Capacity<br>• Forms statements using benchmark fractions and percents<br>• Articulates reasoning for categories<br>• Is able to give examples of consumers for the data | | | | |
| Use of Tools<br>• Uses cards to help organize thinking | | | | |
| Background Knowledge<br>• Draws on knowledge of surveys, nutrition, circle graphs | | | | |

## Rationale

This lesson takes students through the steps of creating a graph: collecting and organizing the information. As students categorize and recategorize their data, they present different stories, suitable for different audiences.

Keep the focus on the data, grouping and regrouping it. Most students need practice generalizing from specifics. The activity of creating category names is useful not only for organizing data, but also for organizing ideas in expository writing. Seeing how a number breaks down into about half, more than half, or less than half is important, but spending time on reducing fractions or converting fractions to decimals and percents is *not* the intent.

Many adults have an intuitive understanding of sample. Sometimes students get sidetracked by the issue (see Question 3, Student 3, below). When students are not distracted by the question, they have a variety of ideas about mechanisms for data collection (from *Practice: Take a Sample*, *Student Book*, p. 24).

2. The Caddy Car Company wants to know whether their customers are happy with the service department. How could they find out this information?

Student 1:
By filling out a customer's check list, to see how many people satisfy with their service.

Student 2:
Well the caddy car company could put a customer service survey about their services.

Student 3:
1. Make a anonymous survey for the customers.
2. Give them the survey during every purchause, if they choose to fill it out.
3. Keep a survey box buy the exit door. (drop in box)

3. Teachers in an adult education program want to know the reasons some students don't stay in the program long enough to complete their goals. How could they get this information?

Student 1:
The teachers need to do a survey on the students in their class to se what's the problem about not staying long in the program.

Student 2:
Assign a essay for the current student attending the program about how some student drop out of high school or for some that drop out of education programs long enough to complete their goals.

Student 3:
May be they having a problem with what they are teaching, or something is wrong with the program, why they did not stay to complete these goals.

## Math Background

Categorical data describe information in mostly non-numerical terms—e.g., ice cream flavors (chocolate, vanilla, strawberry); countries where clothes are made; and foods we eat. Once data are sorted and grouped into categories, we can make numerical statements about them: e.g., about $\frac{1}{2}$ of the people sampled prefer chocolate ice cream, or $\frac{3}{4}$ of our clothes came from China.

The purpose of categorizing is both to show information and summarize it. When deciding upon categories for the data, there are important considerations to keep in mind: how many categories to use and clarity about which data points go into which category. Though two categories work for some information (e.g., yes-no questions: like/dislike; student/not student), they are too few for others (e.g., countries where clothes were made). Having too many categories is not useful either. Sometimes a category with just one data point is included (e.g., only one piece of clothing from Europe).

When it is unclear which category a data point belongs in, perhaps the categories are not well enough distinguished. In foods, for example, if the categories include "grains" and "breads," there might be a question about where to place "seven-grain bread." Practice with sorting and deciding upon categories helps.

## Context

Collecting data is an important part of the process of categorizing data. Foods people eat are interesting because they relate to students' lives and important because conversations about nutrition can arise when the data are shared.

## Facilitation

The more groups of students you have, the longer the debriefing will take. If you can, keep the number of groups to four or fewer.

### Making the Lesson Easier

If students have little familiarity with fractions and percent equivalents, you may choose to compare size only, using terms like "greater," "fewest," or "less than." Or you may create fractions without comparing them to benchmarks or their percent equivalents. Students will see circle graphs again in *Lesson 4.* If you do choose to forge ahead, some ideas for providing scaffolding follow.

Model one example of how you would compare the two categories of data. Find the total number of items for both categories and figure out half of that. For example, consider the total for our raw and cooked food categories which is 17 (an odd number). Think about even numbers on either side of 17 (16 and 18). Half of 16 is 8, and half of 18 is 9, so the halfway point for 17 is between 8 and 9. Then create a fraction: $\frac{5}{17}$ of the foods are raw. The fraction $\frac{5}{17}$ is less than half since 5 is less than 8 or 9. The fraction $\frac{12}{17}$ is more than half. Once you have walked students through the steps, choose another set of data and ask a student to compare the two groups using percents or fractions.

Another strategy for providing scaffolding is to post a chart similar to the one in the *Student Book*, p. 28. Depending on your students' familiarity with fractions, decimals, and percents, leave blanks and engage them in completing the chart.

### Making the Lesson Harder

If your students can handle benchmark fractions and percents, and need a challenge, begin to get them to look critically at the data, including the source and sample size. You might ask:

**If you asked another class to tell you what their favorite foods were, what do you think would happen to the categories? What if you asked the entire community?**

**How do you think your data would compare to data created by another class of adult students in another community?**

## LESSON 2 IN ACTION

*One teacher reports on her observation of students as they created categories together.*

We began with a lively (and hard-to-end) discussion about who would be interested in data about what people eat. I then told them that they were each to write down the three foods they eat most often. I told them not to list drinks and to be specific ("broccoli," not "vegetables"). Some students got this immediately and were visibly exasperated by the ones who kept asking questions ("Can I say dairy, or do I say cheese?" "Cooked food or raw food?") If I hadn't set the limits myself, we'd never have gotten anywhere.

I then took a deep breath and explained that each food needed to be written three times on separate Post-it Notes. I made a set myself and stuck them to the board. Then I collected them and redistributed them to the two pairs and one trio of students. There was confusion for a while but not the total chaos I feared. It really did not take long for them to settle down to lively discussions of how to arrange the data. When students asked me how to organize the data ("Do corn and potatoes go under starch or vegetables?"), I called the class's attention to the question to point out that there are numerous possibilities.

Once the initial confusion was past, students were very involved and discussions were lively, and at times involved more than one group. Everyone had something to say in answer to Janet's question, "What do I call the category that has the cake and butter?" With just a bit of prodding, students were able to use fractions in describing the information. ("Four-ninths of the starch group is rice" was a typical response.) We talked a little about which of the three charts made it easiest to compare categories. We were able to finish the entire lesson, including a decent debriefing.

*Karen Johnsen*
*The Adult Learning Center, College of Staten Island, NY*

FACILITATING LESSON 3

# Displaying Data in a New Way

*How do bar graphs compare with circle graphs?*

## Synopsis

Students have worked with frequency graphs as a way to organize data. In this lesson frequency graphs are the jumping-off point for constructing bar and circle graphs.

1. Students transform frequency graphs into bar graphs and circle graphs.
2. Students post their work, and the class contrasts the two formats, answering the questions: How are the stories the two types of graphs tell similar? How do the stories differ? When do you think it is most helpful to use a bar graph? A circle graph?

## Objectives

- Create bar graphs
- Determine vertical axis intervals for bar graphs
- Construct a circle graph from a bar graph
- Compare and contrast bar graph and circle graph formats

## Materials/Prep

- Colored markers
- Easel paper
- Rulers
- Two copies of a frequency graph on sq. cm grid paper for each student
- Scissors
- Tape

Choose one student-made frequency graph with three categories or more from *Lesson 1, Countries in our Closets* or *Lesson 2, Most of Us Eat ...* . Transfer the frequency graph to a large grid, taking up as much of the paper as the data allows. The larger the size of the frequency graph, the easier the transformation and the more striking the bar and circle graphs will be. After completing Activities 1 and 2, consider extending the learning with a digital exploration similar to these tasks but with a different data collection topic: https://teacher.desmos.com/activitybuilder/custom/60b96512289b625c8308f10e

## Opening Discussion

Until now, the focus has been on frequency graphs. Remind students that in their packet of graphs from the *Initial Assessment*, there were no frequency graphs. Although a frequency graph is a way to organize data, most people actually present their data in a bar, line, or circle graph.

## Activity 1: Constructing a Bar Graph

Demonstrate how the frequency graph can be changed into a **bar graph**. Talk through this process as you do it. Invite students to participate in decision-making where appropriate.

- Draw bars over the x's.
- Color each bar in a different color, making it impossible to recognize individual data points.
- Express mock distress that you can no longer tell how many points are in each category in order to create the need for labels and a $y$-axis.
- Draw a **vertical axis** on the left-hand side of your frequency graph. Label the top of the line $y$.

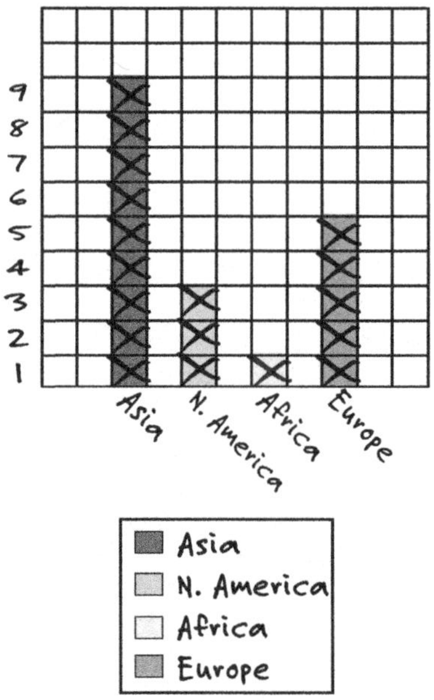


Adding a $y$-axis and labeling it is just one way to solve the problem of bars obscuring the individual data points. Another way is to label the number of items over each bar instead of on the axis. Another way is to simply label the number of items over each bar. You will either have to recall the exact number or refer to a copy of your frequency graph.

## Heads Up!

The emphasis here is on the addition of a vertical axis to the graph and the use of regularly spaced intervals to represent numbers of data frequencies.

Distribute frequency graphs and grid paper to the students, and give them a few minutes to create two identical bar graphs. They need two because they will be cutting one of them up later in the lesson.

## Activity 2: Constructing a Circle Graph

**Circle graphs** are another way to display data.

Demonstrate how the bar graph can be changed into a circle graph. Talk through this process as you do it. Where appropriate, invite students to participate in decision-making, follow along, and make predictions.

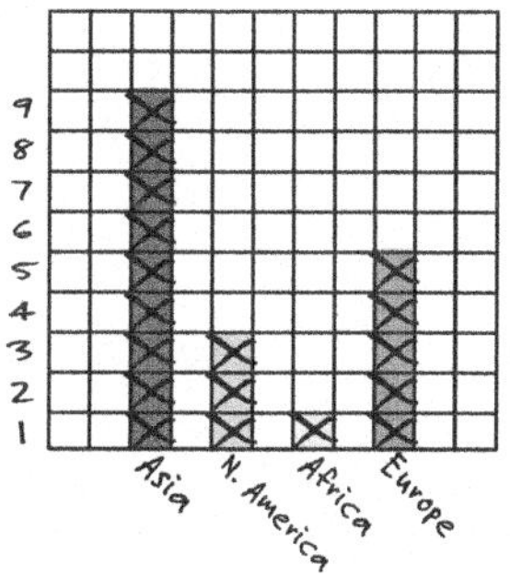


1. Color each bar a different color.
2. Cut out the bars.
3. Tape the bars together, end-to-end, creating one long strip.
4. Create a circle from the strip by connecting the two ends.
5. Trace the circle onto a new piece of paper.
6. Mark off where each category begins and ends.
7. Draw a line from the center of the circle to the marks on the edge of the circle, creating pie slices for each different section.

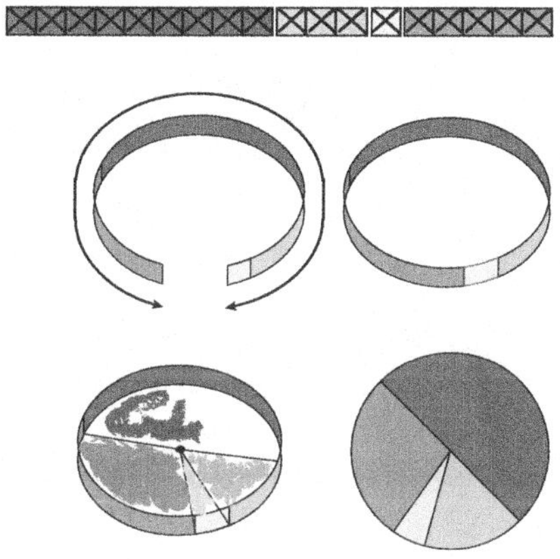

8. To keep track of the data, establish a **legend** or **key**, using one color for each bar. Record the legend next to the circle.
9. Write a title for your circle graph.

## Summary Discussion

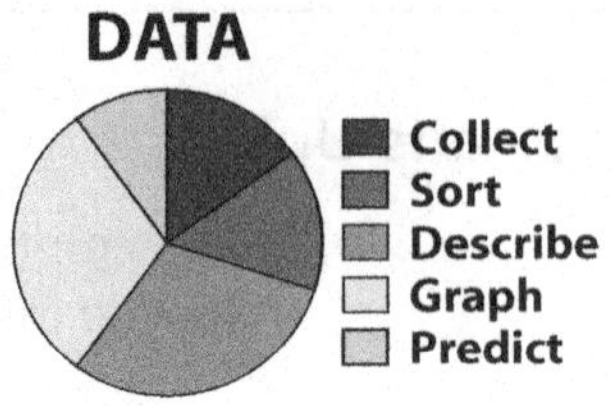


Display all the graphs. Ask:

**How do your bar graphs differ from your frequency graphs?**

**How are they similar?**

**When might it be easier to use a bar graph? A frequency graph? A circle graph?**

**Which graph gives you information you can most easily understand?**

**With which graph is it easiest to compare amounts in fractions? In percents? Why do you think that is? [Push students to give examples.]**

**What fractional/percentage parts jump out at you? From which graphs?**

Give examples:

**I can see fractions more easily in this one…**

Remind students that circle graphs are used only when you are interested in seeing how a whole amount (e.g., an area or a budget) is split into parts. More on circle graphs will follow in *Lesson 4*.

Direct students to *Reflections* questions (*Student Book*, p. 146) for a chance to list new information they learned and the opportunity to pose an "I wonder" question.

## Practice

*Label That Y-Axis,* p. 38
For practice determining intervals on the *y*-axis.

*Bar-Graph Details,* p. 39
For practice labeling the *y*-axis using appropriate scales and reading bar graphs.

*Bar Graph to Circle Graph,* p. 41
For practice creating circle graphs from bar graphs.

## Extension

*Foods You Eat in a Week,* p. 42
Walk students through the steps to produce a simple report with the data they have collected. Remind them of the project's due date. Brainstorm ways to organize the data.

*Net Gains and Losses,* p. 43
For practice exploring negative values and their meanings in a bar graph.

## Test Practice

*Test Practice,* p. 45

## Looking Closely

Observe whether students are able to

### Create bar graphs

Notice who excels at or is enthusiastic about graphs in this class. For those who have motor coordination problems, suggest pairing up. It often helps to have an extra pair of hands to manage the strips of paper, tape, and tracing.

### Determine vertical axis intervals for bar graphs

Notice who is comfortable letting go of counting by ones and who has trouble switching to larger intervals. Provide practice for students who have difficulty with skip counting and labeling the $y$-axis.

### Construct a circle graph from a bar graph

Suggest that students fold the circle graph in half and then in half again, just to ensure that they *do* have the fourths and halves very clearly marked to use as benchmarks. Encourage them to make the first fold along one of the marked lines on their circle graph. In the next lesson, students will have an opportunity to further investigate how to create and interpret circle graphs.

### Compare and contrast bar graph and circle graph formats

Students can get so involved in the colors and cutting that they lose sight of the fact that circle and bar graphs convey the same information in different formats. Pose questions that encourage them to look at the relationships among the data. Bar graphs and circle graphs lend themselves to different comparisons of the data represented. While both formats involve area, in circle graphs the total area is contained within the circle. This allows for very explicit area comparisons among categories in the part-whole format. In bar graphs the area is distributed among the bars, and mental visioning is required for comparisons.

| WHAT TO LOOK FOR IN *LESSON 3* | WHO STANDS OUT? (LIST STUDENTS' INITIALS) | | | NOTES FOR NEXT STEPS |
|---|---|---|---|---|
| | STRONG | ADEQUATE | NEEDS WORK | |
| **Concept Development**<br>• Constructs bar and circle graphs<br>• Compares and contrasts formats<br>• Integrates new ideas, e.g., legend, bars, and labeling *y*-axis<br>• Keeps meaning intact throughout<br>• Labels *y*-axis using appropriate intervals | | | | |
| **Expressive Capacity**<br>• Forms statements with fractions and percents<br>• Comments on the effect of different arrangements of data | | | | |
| **Use of Tools**<br>• Handles grid paper and marking x's<br>• Requests or invents more precise tools, such as compass or protractor | | | | |
| **Background Knowledge**<br>• Refers to other graphs in classroom (from *Initial Assessment*) or other experience with reading and interpreting line and bar graphs | | | | |

## Rationale

Interpreting a graph becomes easier for many people if they have first had hands-on experience constructing one. They are then better able to analyze the parts of someone else's graph. Bar graphs are easier to use than frequency graphs when there are large quantities of data. No one wants to make or count hundreds of x's. The point of having students create frequency graphs in the previous two lessons was to have them understand where data for bar and circle graphs (and later line graphs) come from and how a data set is transformed into a graph.

## Math Background

This lesson brings up the question: Is the format for any set of data interchangeable? See *Lesson Commentary: Opening the Unit,* p. 8, for a review of formats and their purposes.

Often people connect the tops of the bars on a bar graph to help their eyes track the differences between categories. However, when the bars represent data that has been grouped by some characteristic (for example, type of food, type of job, or countries in North America, as in this lesson), it makes little sense to connect the bars with a line because the data here are not connected in a continuous fashion implied by a line graph. Resist connecting the bars in your graph with a line because students might think the line has meaning when in fact it does not. If you changed the order of the categories along the $x$-axis, the slopes of the lines between the categories would suddenly be totally different. The line has no meaning for these activities.

## Context

Circle graphs and bar graphs abound in media: websites, news, social media, and print publications. Data about ethnic composition of people in a variety of occupations, illnesses, and social and economic compositions are available on various websites.

## Facilitation

Some students object to the rough outline of the circle. Many have trouble locating the center, whether their circle is misshapen or not. Address these worries in the way that feels most appropriate for your class. For example, you may want to use one of these strategies:

1. Simply reassure students. The shape of the circle is not the main point; it is not meant to be perfect.
2. Provide tools. Have lots of circular objects available, such as lids of different sizes, so students can use ones that fit their needs. They may also appreciate having a compass on hand.
3. Demonstrate how students can locate the center of the circle by folding it in half, then quarters, after you cut it out. Where the folds meet is the center.

If students tape their circle graphs onto other pieces of paper, they can add legends to their graphs. This will make their final products look more professional.

### Making the Lesson Easier

For those students who do not see how the individual pieces of data from a frequency graph are the same pieces that form the bars of a bar graph, try this: Suggest students draw their x's on graph paper and then color over the x's with a light orange or yellow colored pencil to create a bar. The students can then clearly see both the x's and the bars at the same time.

### Making the Lesson Harder

Give students protractors and compasses. Ask them to construct circle graphs using these tools. Then ask students to label the slices of pie with degrees based on angle size. For example, if one of the bars represented a quarter of the data, the slice of the circle should measure 90°, or 25% of the circle. Be sure to review vocabulary related to circles: "diameter," "radius," "360°."

*Balancing the logistics and the mathematical content, the teacher makes sure her students attend to both. Observed by Martha Merson*

Coordination is a stumbling block. The strips are skinny and do not stay in a perfect circle shape. In fact, I think the thinner the strip, the more often it tips over on its side. Liza draws her circle three times. She informs me that cutting before marking the amount for each slice is not good. She finds tracing on the outside works better than tracing on the inside. Liza is invested in the look of her final product.

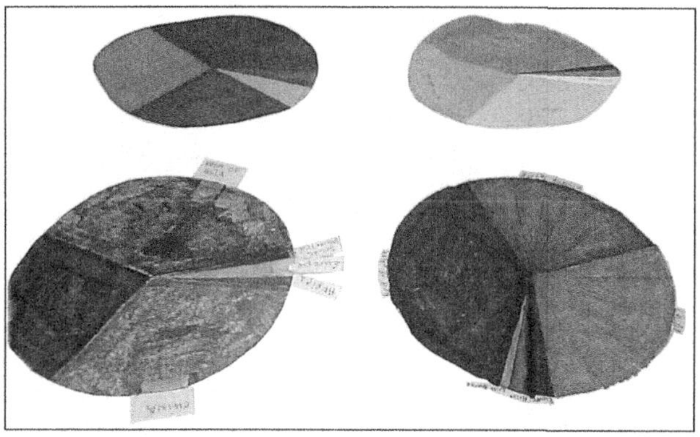

Paul's graph looks good. He makes his circle using a diameter of 8.5″. He locates the center and rotates his ruler around the center to draw eight lines, each 8.5″ long. The endpoints of each line provide him with 16 dots. When he connects them, the circle is very even. When Barbara sees the number of slices, though, she seems worried that the lines will be distracting. But I think Paul could just color over them; or if he does not like the effect, he could cut the circle out and turn it over. I encourage him to keep going rather than start over.

Once the circle graphs are up, they are quite striking. While students admire their work, Barbara asks, "What do you think?"

Theresa says, "Interesting. I learned something. This is the circle graph. Are we doing the triangle graph next?"

Barbara poses another question, "What is your opinion about which one is the best [for presenting the information]?"

A few people comment. I hear Irena say, "Bar graphs. It's instantaneous."

*Barbara Goodridge*
*Lowell Adult Education Center, Lowell, MA*

FACILITATING LESSON **4**

# A Closer Look at Circle Graphs

***How can you tell what percent of the data are in each slice?***

## Synopsis

Students have displayed data using frequency, bar, and circle graphs. They will now focus on the percents pictured in circle graphs.

1. Using information from small surveys of the class, students organize data in tables that show the number in each category, the relative frequencies (fractions or percents), and the total.
2. Volunteers show how they estimate the size of "slices" of a pie chart and collectively they agree on the circle graphs.
3. Students create manipulatives and make a cheat sheet of benchmark percents.
4. The whole class discusses how the slices relate to the data parts and the whole, and how to interpret their meaning.

## Heads Up!

The lesson assumes students can use the benchmark fractions "half" and "a quarter" and their percent equivalents to estimate other quantities.

## Objectives

- Construct circle graphs to show how parts of the data set relate to the total
- Apply an understanding of benchmark percents to estimate the size of "slices of the pie"
- Interpret circle graphs

## Materials/Prep

- Circle graphs from *Initial Assessment* or from *Lesson 3*
- Colored markers
- Paper plates with a midpoint for students to use to check the size of their slices

## Opening Discussion

Begin by reviewing what students know or recall about circle graphs, using their work from *Lesson 3* as a point of reference to evaluate their understanding. Ask:

**What were the steps you took to make circle graphs?**

**How did you know where to draw the lines?**

**What is the difference between a bar graph and a circle graph?**

If no one mentions it, note that turning the bar graph into a circle graph clarified how the size of each category related to the total data set.

Explain that *Activities 1* and *2* of this lesson will help students build familiarity with benchmark percents and circle graphs so they can represent data without having to rely on bar graphs.

## Activity 1: How Many of Each?

This activity walks students through the process of making a circle graph based on survey data and benchmark percents. Students become comfortable and independent with the process as they

- Establish the number in each category;
- Establish the total;
- Relate the numbers to ½ (more than, less than, or equal to);
- Estimate a percent; and
- Check estimates against the total.

### Quick surveys

Take a quick class survey on a question that will yield only two categories, e.g., smokers vs. non-smokers; plan to vote in the election or not; pet owner or no pets; like pizza with pineapple or not; prefer hot drinks or cold drinks; prefer books or movies, etc.

Refer students to *How Many of Each?* (*Student Book*, p. 48), question 1, and ask them to write the category names and number in each category on the table. Clarify the meaning of **benchmark** (e.g., near 25%, 50%, or 75%, or the equivalent fractions). To verify accuracy, ask about the total and compare that to the number of people in the room.

**What fraction of the class voted for [the first category]? Is that more or less than half of the total in the class? How do you know?**

**What benchmark percent would you use to describe the votes for [the first category]? How do you know?**

Ask the same questions about the number of votes for the second category. End with

**What is the result of adding the percents for each category? (100%) Will that always be the case? Why?**

Students will sketch a circle graph. Ask for two volunteers to sketch where the class can see while all students use either the actual numbers from the "How Many" column in *Activity 1* or percents to sketch their circle graph. When they finish, ask the class:

**Is this volunteer's graph a reasonable representation of the data? How do you know it is (or is not)?**

Repeat the question for the second volunteer's graph.

**How are the graphs the same? ... different? How do they differ from yours?**

Then ask several students to explain their processes:

**How did you decide the size of the slices?**

**What numbers or percents would describe these two slices? How do you know? Can you use a number like 5 as well as a number like 25%?**

---

## Heads Up!

Note that there is more than one right way to estimate the size of the slices. However, those students who worked one slice at a time around the graph will have a hard time creating slices of equal size. Encourage them to try a new strategy. See *Looking Closely*, p. 55, for ideas. Emphasize the idea that the circle, which represents *the whole*, is 100%, and the slices are *the parts*.

---

Students complete a second survey and a graph with two categories.

Optional: Instead of asking all students to create a circle graph for the second poll, invite some students to choose to create a bar graph using the same data. Remind students to label the axes and the graph. Have students compare and contrast their graphs.

## Activity 2: Cheat Sheet

To further explore percents displayed in circle graphs, distribute paper plates (see *Facilitation*, p. 58 for using results close to a third, and p. 59 for an alternative visual to paper plates). Save the paper plates. It will be helpful to have these manipulatives available for *Activity 3* and *Practices.*

- Ask students to fold the plates to show 50% and 25%.
- Question what percent would be halfway between 0% and 25%.
- Establish the percent halfway between 25% and 50%.
- Students make dotted lines to show 12.5% and 37.5%.
- Distribute a second paper plate or circle to students.
- Students label the percents in multiples of 10 (10%, 20%, 30%, etc.), based on what they have just learned.
- For each circle, explain how the percents are related to the benchmark fractions (less than ¼, a little more than ¼, etc.).
- Emphasize that the percents are the areas shaded, not the line per se.

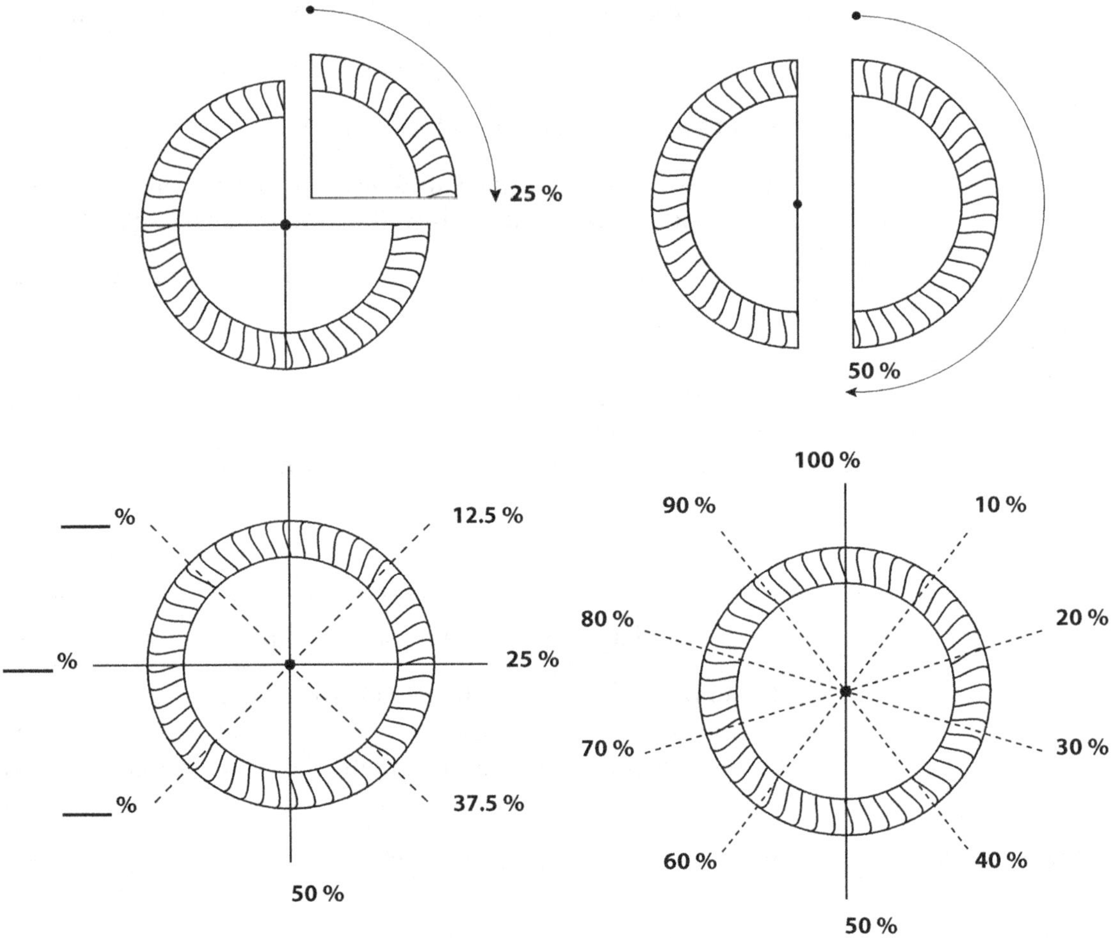


Have students complete *Activity 2: Cheat Sheet* (*Student Book*, p. 50) to reinforce their visual sense of benchmark percents.

Ask students:

**How can you use this information as a cheat sheet to help you?**

**How could the paper plate tool help you if we had (reference a poll in Activity 1) 90% of the votes for 1. ______ and 10% of the votes for 2. ______?**

**How could the tool help if you knew 10 out of 24 students voted for 1.______?**

## Activity 3: Three Possible Answers

Conduct a quick poll offering three possible answers. Use questions such as

- Do you prefer wearing neutral colors, bright colors, or do you wear both?
- Do you work a paid job full-time, part-time, or not at all?
- This week will you mostly cook from scratch, cook from a box, or eat out?
- Do you drink coffee every day, sometimes, or never?

### Heads Up!

Limit the number of options to three. You may add your own vote, especially if doing so will yield friendlier fractions.

Refer students to *Three Possible Answers* (*Student Book*, p. 51), Problems 1 and 2. Pair students so they can compare their sketches and revise them if necessary. Meanwhile, have one student put his or her sketch on the board. Ask students to check the accuracy of their graphs against a paper plate. Remind them to label the slices with percents and whole numbers. Repeat these steps with the second question and survey.

Point out strategies, for example:

- Find benchmark fractions.
- Start with recognizable slices, e.g., ½ or ¼.
- Start with the largest category.
- Total the slices to be sure they equal 100%.

To practice the relationship between fractions and percents, ask:

**If a newspaper were going to publish a headline with the results of the poll, what fractions or percents would they use? (For example, if the number is between 25% and 50%, a sensational newspaper might round up and say, "nearly 50%.")**

Ask students:

**If this graph appeared in the newspaper with no information about the sample, what impression could the data give?**

**How could this graph be used to influence others?**

Check that students are not over-generalizing by asking:

**What was the total for the circle graph we did as a group?**

**Will the total number always be the same?**

**How can you find out what the total is? Where do you look to find the total number?**

**Would it change anything if the circle were twice as big or half as big?**

Because circle graphs are usually represented in percents, sometimes people misread them and believe that the total number of data points is 100. Ask students to turn to Problem 3 of *Three Possible Answers* (*Student Book*, p. 52). The totals are given, but not the parts. However, if students recognize the slices by size, they can easily answer the questions.

Ask pairs to compare their answers with those of other pairs. Then ask:

**How did you decide which town's "yes" count was the highest?**

**How did you find the percents?**

**How did you determine the actual number who voted "yes"?**

Remind students how important it is to know where the data come from, as well as to look at the parts and the whole before jumping to conclusions.

## Summary Discussion

Review what students accomplished: They estimated the size of slices based on benchmark percents, described data with fractions, and sketched circle graphs.

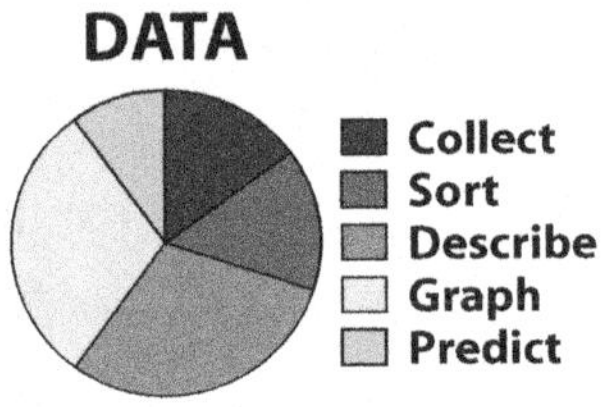


Ask students to list the steps in making a circle graph:

- Find the total.
- Write a fraction for each category.
- Look for a benchmark fraction and/or percent that is close to the fraction for each category.
- Shade each slice, starting with the largest portion.
- Make sure the slices add up to 100% or 1.

Then direct students to *Reflections* (*Student Book*, p. 146), where they make personal lists of what they have learned and need to think about when they encounter circle graphs.

## Practice

*Estimating Percents and Parts of the Whole,* p. 54

*Serve Up the Pie (Graphs),* p. 55
For practice creating circle graphs based on raw data represented in bar graphs. Students first figure out the benchmark fractions, then change them to percents and sketch rough circle graphs.

*Making Predictions,* p. 58

## Extension

*The Computer Makes the Graph,* p. 59
Reviews converting bar graphs to circle graphs with an option to use Microsoft® Excel or Google Sheets.

*Search for Circle Graphs,* p. 60
Students find and interpret their own circle graphs.

## Test Practice

*Test Practice,* p. 61

## Looking Closely

Observe whether students are able to

### Construct circle graphs to show how parts of the data set relate to the total

Watch how students begin the task. Does anyone begin by bisecting the circle? Recommend that students who draw one slice at a time divide the whole into halves first. Do they use all the space inside the circle?

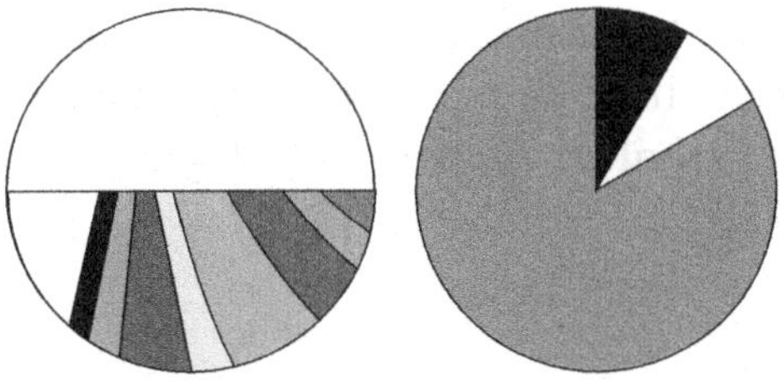

*Some students work with only a portion of the circle as if it were the whole. In the graphs pictured, students left the white sections untouched.*

Offer a checklist that includes at least the following items:

- The circle has a center.
- All lines for slices touch the center.
- Use all the space for the data you have.

Listen and watch for indications that students understand that circle graphs are used to show the categories in relation to the whole. If students refer to the total, or if their language indicates that they equate 100% with the total data set, call on these students to explain their process.

Because circle graphs can represent a small sample as well as a huge sample, students might think that the circle size will change as the sample grows or shrinks. Check for understanding by asking whether the size of the circle matters. Ask whether the size of the slices would change if a greater number of people were polled but the general breakdown in responses remained the same. If students seem muddled, off-track, or unsure, follow up with a new survey. This time distribute circles of different sizes as a basis for the circle graphs. Then give students data from a hypothetical class that is proportional to data from your class. If students say the graphs will be the same, or if they draw the graphs and then reach that conclusion, you will know they understand the relative nature of percents and circle graphs.

### Apply an understanding of benchmark percents to estimate the size of "slices of the pie"

Can students apply their understanding of benchmark percents and fractions? They should be able to estimate what constitutes half, one-quarter, or three-quarters of the data. By comparing the number of items in a category to the total number of items in the data set, they should also be able to describe how that category might be represented as a slice of the "pie." For example, for a category of 4 items out of 60, students should be able to say, "This category will be a very small slice since 4 out of 60 is a small fraction."

### Interpret circle graphs

Can students readily identify slices as being equal to or approximately 25%, 50%, or 75%? Encourage students to compare slices using benchmark percents. Draw their attention to any accompanying information that identifies the total number represented. These two strategies will strengthen students' circle graph reading skills.

| WHAT TO LOOK FOR IN *LESSON 4* | WHO STANDS OUT? (LIST STUDENTS' INITIALS) | | | NOTES FOR NEXT STEPS |
|---|---|---|---|---|
| | STRONG | ADEQUATE | NEEDS WORK | |
| Concept Development<br>• Recognizes benchmark percents based on slices of a circle<br>• Knows percent equivalents for benchmark fractions<br>• Understands that circle graphs show portions out of the total | | | | |
| Expressive Capacity<br>• Is able to explain the relationship between the size of the sample and the size of the slices | | | | |
| Use of Tools<br>• Uses the size of the slices to estimate percents<br>• Uses different tools creatively to sketch slices | | | | |
| Background Knowledge<br>• Uses understanding of percents in creating circle graphs | | | | |

## Rationale

Data are often presented in the form of circle graphs, sometimes simply to inform and other times to influence. Students need to be aware of how circle graphs are created so they can think critically about data shown in graphs.

They also should be able to look quickly at a circle graph and make some general statements, such as which category, if any, represents over 50% of the data.

## Math Background

Although spreadsheet and graphing computer programs allow users to create circle graphs from the same data used to create bar and line graphs, there are situations when one graph is more appropriate than another. Circle graphs (or pie graphs or pie charts) are best used to show how all of the parts of the data relate to the whole. Only the circle graph clearly shows what portion of the total each category is. Circle graphs are not effective in illustrating change over time.

Be sure students understand that the percents represented by each slice are not the same as the actual number of data items in the total (this would only be the case if there were 100 data items in the total amount). Savvy consumers of data check for indications of total sample size. If there are only a few items, each item may account for a large percent. Many adults who have not learned to question the data presented in graphs make erroneous assumptions based on the inference that sample sizes are larger than they really are.

The numbers provided in the *Student Book* total 100%, but if adults look carefully at data reported in newspapers, they may see data that do not total 100%. Much of the time the total will be just under or over 100%. In this case, someone rounded numbers for the different categories, and this affected the total. With surveys, the categories reported may not cover the universe of responses. For example, voting statistics may leave out write-in candidates.

## Context

As in *Lesson 3*, the circle and bar graphs in this lesson illustrate a variety of data. If your class is particularly interested in a topic, collect data for it—or ask your students to collect it—and have them make circle graphs showing the percent in each "slice."

## Facilitation

If you find that the class has created data that involve an amount close to ⅓, you can have them create a paper plate model to use. Suggest that they first divide the plate in half, then fold the two outside edges inward so that they overlap each other evenly. This folds the two halves into thirds. Though this creates six even parts, you can show them how to figure out ⅓ or ⅔ of the total. Encourage them to color and label 33%, and tell students that 33% is rounded.

Alternatively, on a clear plastic sheet, prepare a circle with a dot in the center and lines crossing the center, indicating quarters of the circle. Distribute them. Students draw in other benchmark percents, i.e., 10%, 20%, 30%, 40%, 60%, 70%, 80%, and 90%.

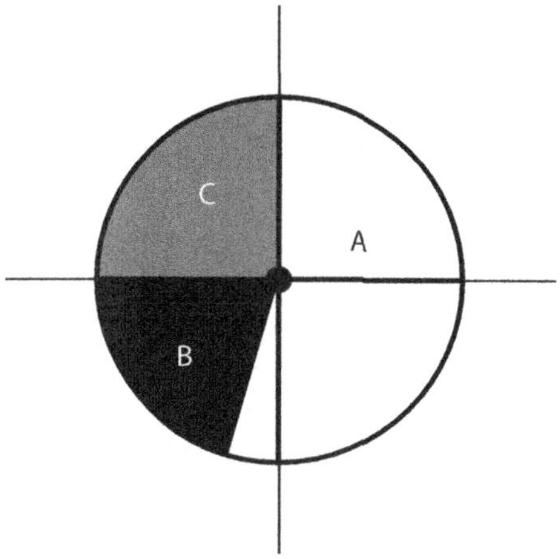


When students need to label slices in circle graphs and are unsure of the benchmark fractions or percents, they can use these transparent circles. By laying the marked, clear plastic sheet on top of their circle graphs and matching the centers, students can use the transparencies as guides to estimate the size of the slices relative to 25%, 50%, and 75%. Model this. Match a transparent circle with the circle from *Activity 1*. If the circle graph from *Activity 1* is bigger than the transparent circle, you will have to extend the lines on the transparent circle. If the circle is smaller than the transparent circle, it should be easy enough to read.

## Making the Lesson Easier

Do a few examples with sample sizes that are easy to manipulate. For example, do a circle graph to show that two out of three doctors recommend patients exercise. Make the numbers increasingly more difficult. The statements below are invented facts:

- Fifteen out of every 20 Americans favor tighter gun-control laws.
- Eighty out of every 100 parents let their kids choose what media to watch with no supervision.

## Making the Lesson Harder

Encourage students who quickly demonstrate the objectives to elaborate on their graphs by adding keys, titles, and descriptive statements. They could also eliminate the label for one slice on their graphs and trade papers, figuring out the missing amount.

*Notice that a difficult concept can be grasped if the numbers are simple and the context familiar. Observed by Lynn McGee*

The teacher, Solange, joins a small group… and the women explain what they are working on. Coaxing them to another level, Solange asks: "In both graphs, do we have the same amount of data to look at?"

"No, it is 34 and 32."

**Solange**: "So the percents are relative?"

The women look confused. "Relative?"

Solange tries another angle. "So the percent means something different in each chart. It is like, if I have four cookies and I eat half, how many did I eat?"

**Student**: "Two."

**Solange**: "If I have 10 cookies and I eat half, how many did I eat?"

"Five," one of the women says, and then the other says, "Oh!"

They are ready to apply this concept, so Solange suggests they edit their written statements.

Later Solange says: "Ideally, I would ask them for an example instead of giving them the example of the cookies."

*Solange Farina*
*BEGIN Managed Programs, NYC*

# Facilitating Interim Assessment: The Data Say

*What information do I need?*

## Synopsis

In *Lessons 1–4*, students organized data, analyzed the results, and constructed their own graphs. This session emphasizes choosing among an existing data set to organize information and answer a question.

### Heads Up!

If the class has completed the *Lesson 3, Extension: Foods You Eat in a Week,* you may skip this *Interim Assessment.* Before continuing with *Lesson 5*, collect the *Foods You Eat in a Week* homework. Post the work with students' permission. Since both you and the students submitting the work should be involved in its assessment (see *What to Look for in Foods We Eat Project* and *Assessing the Foods Project Homework,* pp. 67 and 70), set a time for feedback.

1. The whole class discusses decisions, brainstorms and narrows criteria, and then lists sources of information.
2. Working in pairs, students choose among data.
3. Students arrange the data as a frequency graph, bar graph, or circle graph.
4. Students present the information and answer a question or make a recommendation.

## Objectives

- Alternate between different graphing formats (bar and circle graphs)
- Use either a bar or circle graph to convey information, make decisions and predictions, and pose questions

## Materials/Prep

- Colored markers
- Easel peper
- One copy per student of *Interim Assessment*, pp. 147–153

Post a list of upcoming decisions, drawing from what students listed in their *Initial Assessment*. Then add "Choosing an apartment" and "Choosing a new partner."

The data sets vary in their levels of complexity, so think about which *Activity* options would be most appropriate for which students.

- "Your Survey," (p. 150) suggests students survey 10 people. If they deviate from the questions suggested, you may need to talk with them to ensure that the data will be manageable.
- "Crime Fighter," (p. 150) is a simpler story. The data are reported in whole numbers rather than percents. Students will need to select among data to construct a bar or circle graph.
- "Eating Out," (p. 151) provides data in the same format used in *Lessons 1* and *2*. Students decide, based on the data, where a restaurant should spend its advertising dollars.

Prepare a data set based on your program's enrollment data.

- Review "A Look at Your Program," (p. 149). Have the data on hand or inform the administrative assistant that students might request the data.

Prepare alternative data sets, if you choose.

- For example, supply your own chart of "Foods I Eat in a Week" along with a research question such as "What indications of healthy eating are there? What could this person do to improve his or her diet?"

## Opening Discussion

The *Interim Assessment* is an opportunity for students to synthesize what they have learned from *Lessons 1–4*. Provide the basis for using data in decision-making and choosing among data by saying:

**At the beginning of the unit, I asked you about decisions you will need to make in your life soon. What do these decisions have to do with the work we have been doing?**

Listen to students' comments. If they have not made the connection, point to one of the upcoming decisions on the list you prepared.

**Quickly brainstorm some criteria that might influence your decision in one of these situations.**

Circling one of the criteria on the list, say:

**Data are information. What data could you collect to help make a decision based on these criteria?**

Record ideas, then transition to the activity by saying,

**So, in a real-life situation you might not collect all the possible data. You focus only on the important criteria needed to make an informed decision.**

## Assessment Activity: The Data Say

Give an overview of the activity. To further clarify what each of the groups will need to address in its presentation, see instructions on p. 148.

**You will be given a story and a set of data.**

**Think about how this information could be used, based on the story provided.**

**Then create either a bar or circle graph using some of the data. You will *not* use all of the data when you create your graph. Pick the data that fit your need.**

**Look closely at the graph you have created. What does the graph suggest to you? How do the graph and the data help you address the questions?**

**Prepare a presentation. Share your story, and present your graphs and your response to your question. Be sure to provide reasoning for why you chose the format you did.**

If students wonder why you are asking them to do a presentation, tell them that math, and especially data, are used to communicate and make decisions. They need to be able to use the language of graphs and data to communicate, make decisions, and persuade others. Practicing in class with a "friendly" audience will give students an edge in the job market, since self-presentation plays a role in interviews.

Assign groups to work with each option, and give students a time limit. If needed, reiterate what they are to do with the data.

If students struggle to get started, suggest they underline key words and use them as a basis for categories. Then ask:

**What two or three categories would help you answer the question?**

**Does it make sense to compare these two or three categories?**

---

### Heads Up!

If students choose to construct a circle graph, make sure the categories and data they select make a sensible whole.

---

Review the guidelines for giving the presentation when the class reassembles (see p. 148). Ask for student contributions to the guidelines and record their contributions with the ones on p. 148 where students can refer to them as they plan their presentation.

Before students begin their presentations, provide a strategy for groups to give and receive feedback. For example, you can ask them to complete stems such as, "I liked it when you…" and "I wonder what would have happened if you…" If your students do not handle oral feedback well, you may suggest they provide written feedback, using the criteria they suggested for effective presentations.

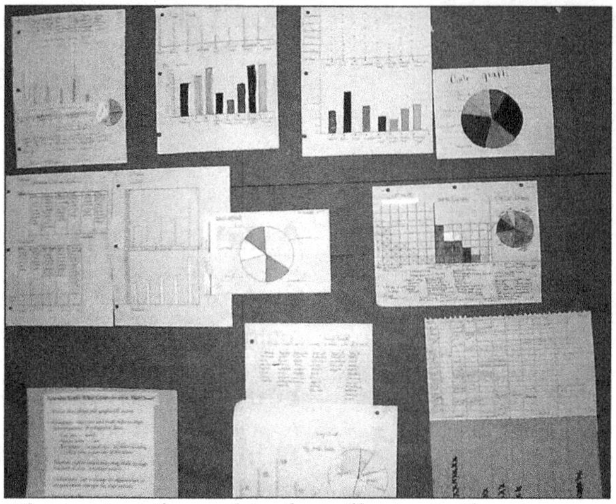

*The posters from this class illustrate the variety of choices students make about graphs.*

Ask the listeners questions such as

**Can someone restate/summarize the point of this graph?**

**How does the graph support the decision made by the presenter?**

**What questions does the graph raise for you?**

## Summary Discussion

Ask about the process of organizing, interpreting, and presenting data. Pose questions such as

**How did you decide where to begin?**

**Was there something challenging for you as you worked with your data?**

**Was there something particularly challenging for you as you planned and did your presentation?**

**Were there data you wished you had?**

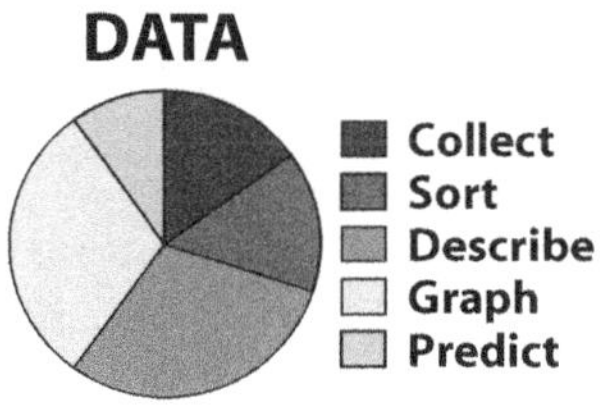


Recap by going over some of the findings articulated by the students. Remind them that though they drew conclusions, they had a limited data set to work with. They should continue to ask themselves what other information would be useful.

To close, refer students to *Reflections* (*Student Book*, p. 147) where they can write about what they learned in this unit.

## Practice

Little practice is provided since this lesson is a chance to reflect and evaluate. If you have an extra story, you may assign it as homework.

*Decisions, Decisions*, p. 152.

## Looking Closely

Observe whether students are able to

### Alternate between different graphing formats (bar and circle graphs)

Do students remember the steps for constructing graphs? If not, prompt them first to look back at their own work in *Lessons 1–4*. If this is not sufficiently helpful, give more specific suggestions.

The students' job is to choose from the information provided and display it in a circle or bar graph. Their presentations must include how the format affects the impression of the information. Are they thinking about the quantity of information; whether it is given in percents or raw numbers; whether it is a total split into parts; whether it is presented in a frequency graph or a bar graph? What reasoning do groups give for deciding on one format over another? Is it based on the nature of the information, the point they want to make, or, more likely, on their own comfort level with the various formats?

### Use either a bar or circle graph to convey information, make decisions and predictions, and pose questions

Is there evidence that students are making sense of the information they have? Are they able to relate to the big picture as well as to the details? Where the graphs include an element of change over time, listen for students' comments on trends. For example, can they articulate that there is change upward or downward and that their recommendation should take this into account?

During the presentations, you have an opportunity to see how students explain their interpretations of their graphs. You will also note how they communicate their understanding to others.

| WHAT TO LOOK FOR IN *INTERIM ASSESSMENT* | WHO STANDS OUT? (LIST STUDENTS' INITIALS) | | | NOTES FOR NEXT STEPS |
|---|---|---|---|---|
| | STRONG | ADEQUATE | NEEDS WORK | |
| Concept Development<br>• Makes decisions about what data to select<br>• Locates and explains the data, axes, labels, and legend<br>• Chooses a format to display the data<br>• Creates a graph from the data | | | | |
| Expressive Capacity<br>• Explains the main idea of the graph<br>• Explains how the decision was reached based on the data<br>• Answers audience questions<br>• Uses statements with mathematical language | | | | |
| Other: Presentation/Team Work<br>• Actively participates in planning and presenting<br>• Includes everyone in preparing for the presentation<br>• Includes all group members in the presentation | | | | |
| Background Knowledge<br>• Has made/participated in presentations<br>• Provides criteria for effective presentation based on prior experience | | | | |

| WHAT TO LOOK FOR IN *FOODS WE EAT PROJECT* | WHO STANDS OUT? (LIST STUDENTS' INITIALS) | | | NOTES FOR NEXT STEPS |
|---|---|---|---|---|
| | STRONG | ADEQUATE | NEEDS WORK | |
| Concept Development<br>• Demonstrates ability to organize information<br>• Creates frequency graph, bar graph, and/or circle graph from raw data | | | | |
| Expressive Capacity<br>• Chooses appropriate titles for groups of data<br>• Makes statements about data using mathematical language<br>• Generalizes or summarizes from data | | | | |
| Use of Tools<br>• Uses tools such as graph paper, ruler, etc. to create an easy-to-read final product | | | | |
| Background Knowledge<br>• Uses understanding of ingredients and nutritional information to categorize data | | | | |

## Rationale

This lesson is designed explicitly to show students how much they have learned. The goal is for students to sort through data to find usable information that addresses a question and, through the presentation, to make a recommendation or recommendations based on the information.

## Math Background

This lesson concludes the section of the book on categorical data. Now, the focus will be on graphs that use numerical data.

## Facilitating

Because this lesson is designed to be a *Interim Assessment*, keep yourself out of the role of directing the work. Record as much evidence as possible for what students have learned and are now able to do. Even if they cannot complete the activity independently, make notes on the types of questions they ask and their ability to communicate effectively with their peers using precise language.

Ideally, you and your students together will list criteria for a successful presentation. Write down all ideas, and then prioritize the list. Following are some ideas along with points to use as a guideline for weighing the various criteria in assessing students.

- Including everyone in the presentation **10 points**
- Being clear about the information in the graphs **30 points**
  - Explaining the horizontal and vertical axes (if a bar graph)
  - Explaining where the numbers come from
  - Explaining the main idea of the graph
- Using statements with math terms (fractions, percents, words such as "double," "twice," "triple") **20 points**
- Keeping the class engaged **10 points**
- Being clear about the next step or recommendation **20 points**
- Answering audience questions clearly **10 points**

Students cold use Microsoft® Excel or Google Sheets to construct their graphs.

## Making the Lesson Easier

Substitute simpler data or provide a very limited set of data to choose from if the options are too difficult. The focus should be on getting students to choose data to create a graph and then make predictions or decisions. In real life, students are continually bombarded with data and need to be able to determine which information is important and why. This lesson will help them build skill in this area.

Ask questions as students ponder which data to use. Encourage them to think out loud about which data should be used and why.

## Making the Lesson Harder

You could substitute data with more complex numbers or a larger data set. You might ask students where they could collect data to answer a question or make a prediction or decision. Learning how to perform an in-depth search provides an opportunity to learn how to judge the value of information.

## Assessing the Foods Project Homework

| ASSESSMENT CATEGORY | SAMPLE TEACHER NOTES/ OBSERVATIONS |
|---|---|
| **Concept Development**<br>• Demonstrates ability to organize information<br>• Creates frequency graph, bar graph, and/or circle graph from raw data | • Categories are meat, dairy, grain, condiments, and junk food.<br>• Accurate. |
| **Expressive Capacity**<br>• Chooses appropriate titles for groups of data<br>• Makes statements about data using mathematical language<br>• Generalizes or summarizes from data | • Poses a question: "What food is most common in Amanda's diet?" but does not make statements.<br>• Check with student regarding ways to strengthen description of results. |
| **Use of Tools**<br>• Uses tools such as graph paper, rulers, etc. to create an easy-to-read final product | • Final product is neat, clear, no erasures, circle slices done with a ruler. |
| **Background Knowledge**<br>• Uses understanding of ingredients and nutritional information to categorize data | • Uses traditional categories such as meat and dairy, and improvises "junk food" to cover sugary and salty snack foods. |
| **Next Steps for this Student** | • Give her sentence stems and ask her to include benchmark fractions or percents. Ask her for one summary statement about the foods she eats. |

# FACILITATING LESSON 5

# Sketch This

***Words can tell a story, but so can lines.***

## Synopsis

The past few lessons have centered on organizing and presenting data. In this lesson, the focus is on using a **line graph** to represent **change over time**.

1. The class works as a whole group, sketching line graphs based on statements in a story they hear.
2. Students have three more opportunities to listen to a story, sketch line graphs to show the change over time, and see how closely the story and their line graphs match.
3. Students develop a vocabulary to use in composing stories to go with graphs in *Lessons 6–10*.
4. The class discusses how labels and titles need to correlate with the stories told in the graphs.

## Objectives

- Sketch a line graph based on a narrative
- Demonstrate an understanding of how the shape of a line provides information about what is happening over time
- Improve precision in describing change over time by adding words to students' working vocabulary

## Materials/Prep

- Colored markers
- A place to create and display about 15 graphs—you will refer back to these student-generated graphs to make a point about labels, titles, and convention when summarizing the activity.

Prepare six drawing spaces for graphs by drawing a large horizontal and vertical axis on each drawing space. Prepare to display each of the following statements separately:

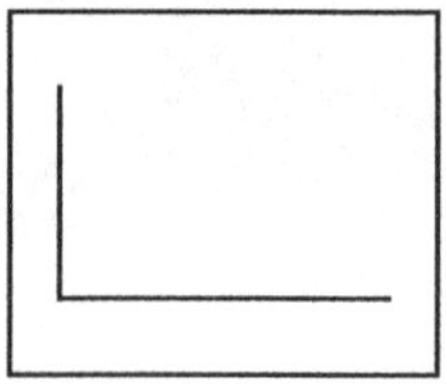

1. "Giselle's baby's temperature rose sharply between 6:00 p.m. and midnight; Giselle gave her baby aspirin, and the baby's temperature dropped rapidly and stayed close to normal for most of the next day. At 6:00 p.m., however, it began to rise again, and Giselle was worried."
2. "Martha ran a marathon. In the first hour and a half, she ran at a steady pace. Then she slowed down for the next hour, saving her energy for a push in the last half hour when she ran faster than at any other time."
3. "Brian followed the share price of a company's stock one Monday, trying to decide whether he should buy stock. He found that in the morning it was at $25 a share; by noon it had plummeted, and by the end of the day it rose slightly."

## Opening Discussion

Begin by referring to past work on bar and circle graphs and introducing an idea you will return to in the summary:

**Over the past few lessons we have focused on bar and circle graphs. What is another type of graph you often see? How do you think it might be used?**

## Activity 1: Sketch This

Follow these steps for each statement:

- Display and read the first statement from Materials/Prep about Giselle's baby's temperature."
- Using the drawing space with the L-shape of the axes already in place, ask for two to four volunteers to draw their sketches for other students to see.
- Point to the first sketch. Ask for new volunteers to trace the lines along the graphs while telling the story. Allow students to make corrections as necessary. Solicit agreement from the other students.
- After students make and discuss their points, reach a consensus.

Read statements on the marathon and the stock price. Each time you read a statement, repeat the steps:

- Two, three, or four students sketch a graph on newsprint.

- Two or three different students volunteer to come up front to follow the lines on the graphs while telling the story and making revisions or corrections.
- Encourage students to discuss their reasoning for the changes throughout this process or afterward.

Before moving on, ask students to make a class vocabulary list from the words and phrases that were used in describing the stories, including but not limited to "rise and drop," "no change," "sharp increase," and "sharp decrease." Make sure students copy this complete list into their student books.

Ask students to consider specifics of the line graphs by pointing to parts of the graphs and asking:

**What do you think is going on in this section here?**

**What does this point tell you? What does it tell you in relation to the points that come before and after it?**

**What does it mean that this line is steeper than this one?**

**If the same pattern or trend continued, what do you think would happen?**

Refer students to *Sketch This* (*Student Book*, p. 64), where they will make a sketch independently.

## Activity 2: Label This

Open a discussion on labeling. In all of the graphs sketched, labels would help a reader make sense of the information. For the river level, the title might be "River Level," and the labels on the horizontal ($x$) axis might correspond to days of rain, with the numbers 1, 2, 3, 4, and 5 equally spaced. The graph designer could also choose to label the $x$-axis with dates or name actual days of the week. In this particular graph, the vertical ($y$) axis does not need a label because the statement does not quantify the water level. If students choose to quantify the water level, make sure their choice of unit and interval makes sense. Units of depth in inches or feet would be logical. Work with the class to label the axes and title the graphs they have made so far.

During this lesson, students might raise questions about the numbering and spacing along the **horizontal axis**. This is an important question. If students do not bring it up, you can ask:

**"Would it change anything if I spread these numbers out or wrote them closer together?"**

For now the students can just speculate. During the next few lessons, they will have a chance to observe how these decisions have an effect on the reader.

At this point, you could also note that the horizontal axes in graphs from past lessons had labels that were names of countries or food types. Now some have labels that are measures of time.

After looking at the four graphs and talking about the titles, you might want to say:

**By convention, graphs of change over time show time on the horizontal axis, so that as we read from left to right, we can see what happened from beginning to end—very much like a story.**

## Summary Discussion

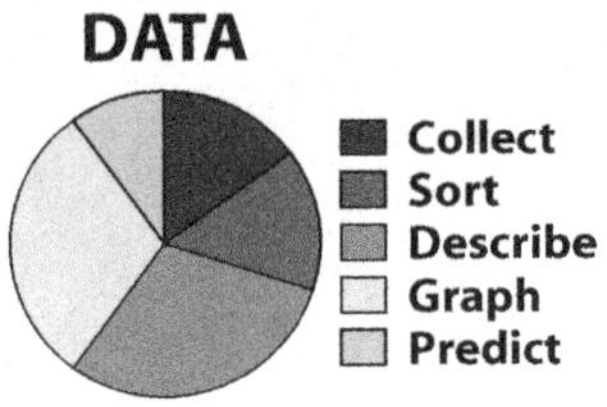


Summarize for students what you have observed them doing well. For example:

**You all paid attention to words like ...**

**You changed the direction of the line.**

**You graphed change over time.**

**You did not get hung up on the numbers on the $y$-axis.**

**You made up reasonable labels for the $x$-axis.**

This will help students isolate and build on the skills important for the next lessons on line graphs.

End by saying:

**This was our first time working with line graphs. Think about line graphs and bar graphs. How are they similar? How are they different?**

Leave students with these questions to think about. There is no need to resolve them yet.

Direct students to *Reflections* (*Student Book*, p. 147), where they can write about how graphs tell stories.

## Practice

*Candy Bars,* p. 65
For practice arranging four graphs from slowest to fastest growth. This one is critical— ideally students complete it before doing *Lesson 6 Roller-Coaster Rides.*

*How Does It Go?* p. 66
For practice sketching a line based on a description of a day of crazy weather.

*Looking at the Entire Graph,* p. 67
For practice telling the story of a complete graph.

*Labeling Graphs,* p. 68

*Telling a Story,* p. 69
For practice inventing a story to correspond to the shape of a graph.

## Extension

*The Shape of a Line from Your Life,* p. 70

*Elevator Traffic,* p. 71

## Test Practice

*Test Practice,* p. 72

## Looking Closely

Observe whether students are able to

### Sketch a line graph based on a narrative

This lesson provides an opportunity to see how students handle line graphs representing change over time. How do they orient their lines? What factors influence their decisions to change direction or slope? Do they know to start the line at the left and move toward the right? Do they start at a sensible point, either zero or higher, depending on the statement? Does the line change according to the narrative?

### Demonstrate an understanding of how the shape of a line provides information about what is happening over time

Students should notice that the incline of the line, along with the labels on the $x$- and $y$-axes, are clues to how fast something is changing over time.

They should also begin to understand that the direction of the slant, if there is one, gives a quick indication about whether there is an increase or decrease.

### Improve precision in describing change over time by adding words to students' working vocabulary

Students should begin to become fluent using terms such as "increase" and "decrease." They should understand and use terms such as "gradual," "sharp up-turn," and other similar expressions that describe change over time.

| WHAT TO LOOK FOR IN *LESSON 5* | WHO STANDS OUT? (LIST STUDENTS' INITIALS) | | | NOTES FOR NEXT STEPS |
|---|---|---|---|---|
| | STRONG | ADEQUATE | NEEDS WORK | |
| Concept Development<br>• Orients the line from left to right<br>• Starts the line at a sensible point<br>• Sketches increases and decreases that correspond to the narrative<br>• Marks intervals or labels with consistent spaces | | | | |
| Expressive Capacity<br>• Articulates reasons for revisions to graph<br>• Defends reasons for the shape of the sketch<br>• Expresses relationships between graph and story | | | | |
| Use of Tools<br>• Works comfortably within the $x$- and $y$-axes | | | | |
| Background Knowledge<br>• Has experiences similar to the ones described in the statements<br>• Refers to past work with line graphs or other types of graphs to support his or her ideas | | | | |

## Rationale

This lesson focuses on change over time, as represented by line graphs. Because this lesson asks students to get involved with line graphs from the beginning of the first activity, it can be a baseline assessment of their ability to represent a statement of change over time using a line graph. Primarily, students gain experience listening to descriptions about something that changes over time and drawing a line graph to represent that change. This lesson also provides an opportunity to discuss labels and titles and the difference between labels on frequency graphs (all places or numbers) and line graphs (all dates or times).

## Math Background

Just as a story has a topic, a beginning, and an end, as well as high and low points, so do line graphs. Students need to pay attention to the title, labels on the axes, and—reading from left to right— the story told about change over time. Increases, decreases, and no changes are the most important effects that they will learn to represent using the line.

Students may sketch small graphs or very large graphs. Samples of work from two different students may show the general upward and downward trends, but the specific interpretations may vary since the size and scale of the different drawings can influence the impression of the story. Scaling is a topic taken up in later lessons.

Labels give context to the graph, telling what is being pictured.

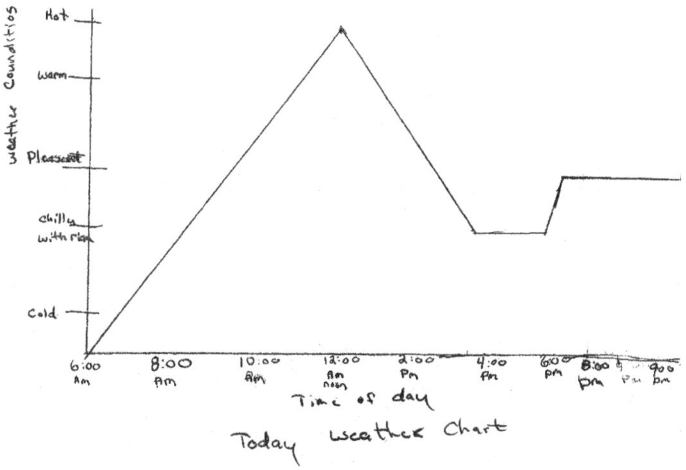


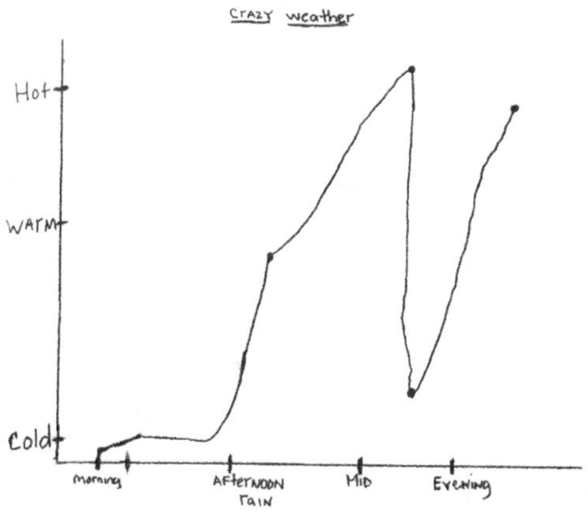

## Context

The stories that students will graph in this lesson are examples of real-life occurrences: a baby's temperature, a marathon run, a stock-market purchase, and a river's rising water level. Other stories you might want to consider graphing could be about the number of people at a grocery store, movie theater, library, or fast-food restaurant over a 24-hour period; the number of swimmers at the beach over a 12-month period in a given city; or the number of buses or trains at a particular stop or station over a week or a day.

## Facilitation

Almost every teacher has ideas about how to modify the facilitation of this activity. Some teachers want students to have printed copies of the statements to facilitate making fair assessments. Some suggest having all the students sketch while seated and then asking one or two volunteers to come to the front of the class; this avoids putting students on the spot. However, another approach, asking volunteers to come forward and sketch on newsprint in response to dictation, has several advantages. Even though this may put students on the spot, it makes the process of translating words to pictures and working with those pictures visible to all. It also generates enthusiasm, interest, and energy.

*The math is more complicated than it seems!*

There are two volunteers at the board. They listen attentively to the statement about the marathon. As they begin drawing, one participant asks, "How long did she run all together?" It is hard to keep track of all the information.

The two pictures start off the same, with a flat, horizontal line, but one dips more gently, the other more sharply. Both end with a diagonal line up toward the right.

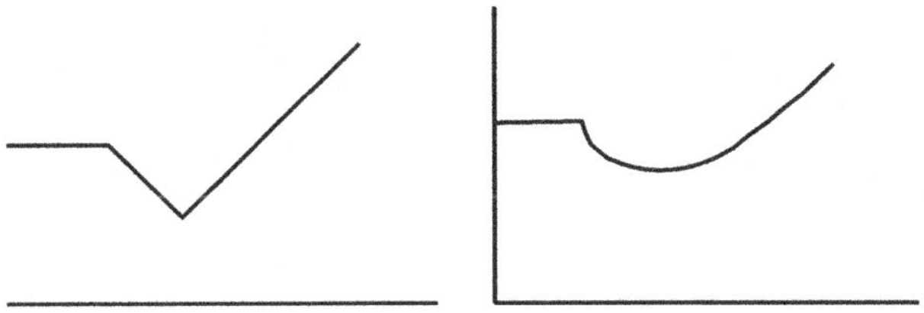

original sketch

The two volunteers sit down and two more get up. Lillian says the only thing she would change is the width of the flat line (in both graphs). The first part should be longer: since the person in the story ran an hour and a half at the same pace, compared to an hour for the second part and a half hour for the third part.

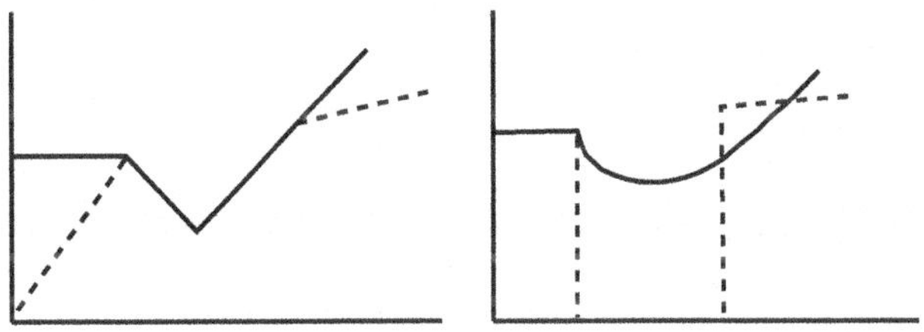

—— original sketch ----- revision

Everyone seems pleased with the lines until Eleanor comes to the front of the room.

She points to the line's origin on the left, at the $y$-axis. "You do not start out running at this speed. You work up to this speed," she says. "When you are starting, you have to start at nothing and work up to the steady pace."

The first volunteer admits, "If you are really going to analyze it, you are right. Maybe what the story should have said is at some point in the race, she was running a steady pace."

*(continued on next page)*

*(continued from previous page)*

Clearly there are two options for revising, changing the drawing or changing the statement.

Mary Jane asks, "What does that bottom spot mean? What is happening there?" The group members begin to reply that the point is a low speed.

Mary Jane asks, "Is that a stopping point?"

The group seems to agree that it is not.

Mary Jane reminds everyone, "These are the questions to ask about line graphs: What do these lines mean and what does a turning point mean?"

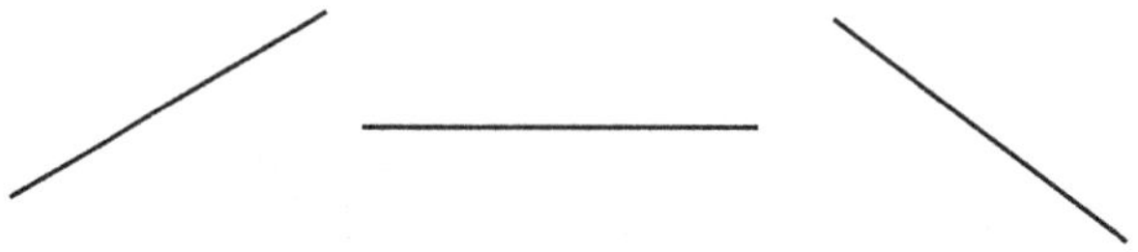

No one asks why the graphs look so different. Are both correct interpretations of the statement? And can a marathon end with increasing speed?

*From a workshop with teachers and assistant teachers*
*New Brunswick, NJ*

# FACILITATING LESSON 6

# Roller-Coaster Rides

*What story does the graph tell?*

## Synopsis

In this lesson, students generate statements about a graph showing change over time. It builds on the previous lesson, in which students sketched graphs based on stories.

1. Together students identify problems in how axes are labeled.
2. The whole class discusses a graph, writes true and false statements, and then carefully analyzes the statements.
3. Students visit stations to gain practice in interpreting line graphs and communicating trends.
4. The whole class debriefs together.

## Objectives

- Use new terms to describe graphs that show change over time
- Isolate and describe pivotal points in a graph of change over time
- Refine ability to sketch line graphs based on oral information
- Distinguish between true and false statements about a line graph

## Materials/Prep

- Prepare to display an enlarged version of the following graph for the whole class. Make sure to space the dates evenly.

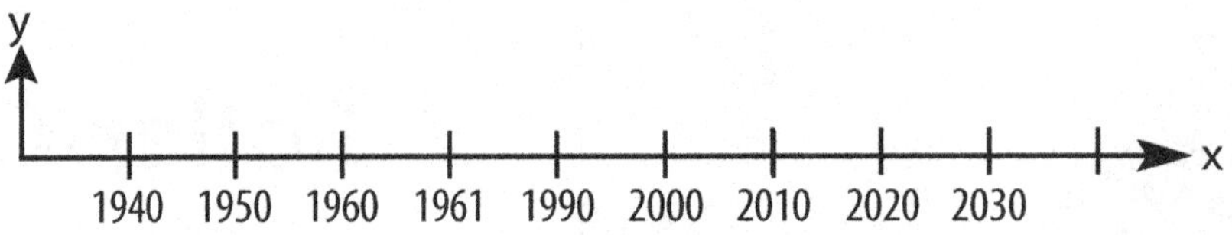


- Display the vocabulary list from *Lesson 5*.

Optional: a display of the graph "Ups and Downs at the Park" (*Student Book*, p. 76)

Prepare stations. For Stations 1 and 2, make copies of graphs (*Blackline Masters 5* and *6*) and graph paper (*Blackline Masters 10–12*); see directions in *Student Book*, p. 77. For Station 3, post the 1099 forms from *Blackline Masters 7* and *8*. For Station 4, post the table "Price of a Subway Ride," *Blackline Master 9*.

## Opening Discussion

Explain that although students sketched graphs in *Lesson 5*, people are more often required to interpret graphs than to construct them. Remind students that the labels they used for line graphs helped them make sense of the sketches. Then display the graph you prepared and say:

**I saw this graph the other day. What is wrong with it? What makes it problematic?**

If students do not catch on, point to the first two dates and ask how many years are between them. Ask the same of the next set of dates. Labels give the **scale**, the size of the space or **interval** between numbers.

After establishing that the increments are not consistent, ask:

**How do you think that might affect the graph? What would be the problem with this?**

If students are unsure, sketch in a line showing a decrease between 1960 and 1961, and ask about the drop:

**Does this look like a steady drop, a gradual drop, or a dramatic drop?**

**If 1960 and 1961 were closer together, would the slant of the line change?**

The important thing for students to realize is that the spacing between numbers does have an impact on a reader's impression of increases and decreases. Students should be aware of this as they construct graphs.

## Activity 1: Amusement Park Attendance

Ask students to look at the graph of annual attendance at amusement parks (*Student Book*, p. 76), and engage the class in making sense of the information it conveys. Build on the language used in the last lesson.

**What are the *trends* you see? What does the line tell?**

Engage students in plotting points by asking:

**How can I include the data for 2020 (317 million) and 2021 (319 million) on the graph?**

If necessary, model finding the date on the $x$-axis and the corresponding data point on the $y$-axis, tracing up and over until you find the right point.

Ask students to describe what they see and make two statements, saying:

**Today's activity is a game called "Two Truths and a Lie." To play, write down two true statements and a false statement about about the Ups and Downs at the Park graph on page 76 (*Student Book*).**

Remind students of the graphs they worked with in the previous lesson. Tell students their statements should describe a change over time, similar to those used to express a river rising, a marathon pace, and a baby's temperature.

---

### Heads Up!

Make sure students understand the graph well enough to begin writing statements. If the whole-class discussion is too thorough, however, the game will seem redundant. Use your judgment in deciding how much time the class spends discussing the graph.

---

Ask for students to contribute statements about the roller-coaster graph until you have about 10. Display them for the whole class to see.

Then without judgment or comment:

- Post and read each one. Ask everyone to vote "True" or "False."
- Keep a tally of the number of students who vote "True" and "False."
- If students' statements are not clear to the whole class, ask them for further clarification.
- Revisit the statements, especially those with mixed votes, and discuss them. The writer of the statement should express his or her intention, add clarification, and be the first to suggest changes.
- Ask students to suggest changes to make the false statements true.

## Activity 2: Stations

Review directions for each of the different stations. Directions for the first two stations (*Ups and Downs*) are similar to those in the previous lesson. Students work in pairs; one describes a graph as the other sketches it, and then they switch roles.

**Station 1**: One partner holds the graph "Approximate Cost of a Share, Pharmacyclics Stock (*Blackline Master 5*) and tells the story of the graph, without showing the graph to his or her partner. The storyteller's instructions are

- Start with "This graph is about..."
- Describe the shape of the lines.
- Use words like "doubled" or "increased by $0.50."
- Give your partner time to sketch the graph based on your story.

The other partner follows these instructions:

- Listen to the story of the graph.
- Sketch the graph, as you did in the last lesson.
- Label and title your sketch.

When the sketch is complete, the pairs

- Look at the graph together.
- Talk about how the sketch and the graph are the same and how they are different.
- Review the description and talk about which statements were difficult to phrase or hard to understand or sketch, and which were clear.
- Make notes on the most helpful words.
- Move to the next station and switch roles.

**Station 2**: Pairs then switch roles and work through another graph ("Average Milk Prices per Gallon, U.S. Cities 2012–2022," *Blackline Master 6*).

---

## Heads Up!

Students should *not* tell their partners how to make the graphs. One partner should just tell the story and let the other partner make the sketch.

---

**Station 3**: *Jerri's Jobs*
Students locate information on nonemployee 1099 forms, plot a graph, and then answer questions about changes in Jerri's annual salary.

**Station 4**: *The Price of a Subway Ride*
Students use data from the table (*Blackline Master 9*) to plot the prices, then draw a dotted line to see the changes in price over time. They describe the changes

in the cost of a subway ride during the 1980s and 1990s. Discuss with students why simply connecting the points will not give a true picture of the price change. The price does not change continuously, but in discrete jumps.

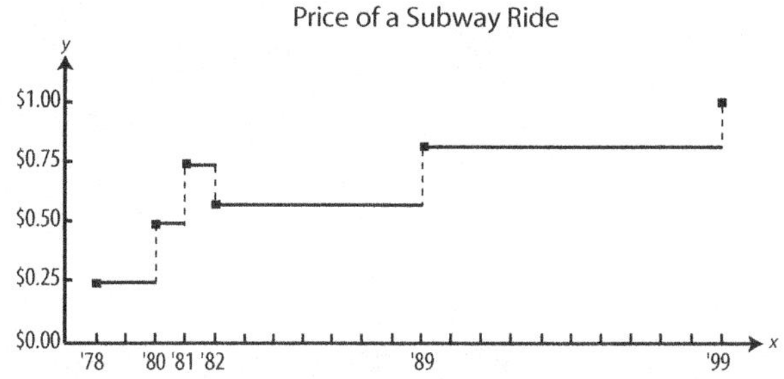


## Summary Discussion

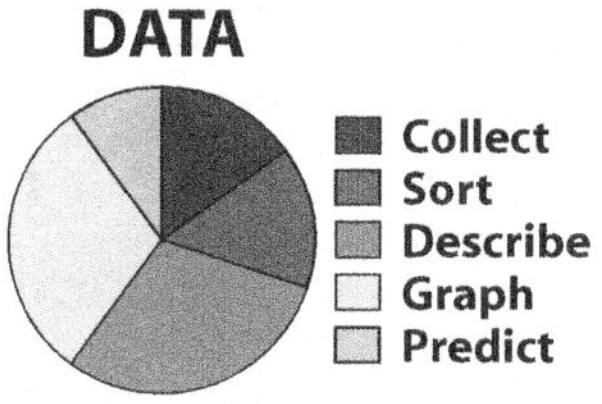


At the end of this stations activity, ask volunteers to share what they learned about describing line graphs. To facilitate this discussion, you might ask:

**What phrases did you use to tell your story? Which phrases were most helpful?**

**How did you show a big change? How did you show no change?**

**What kind of line indicates a steep increase? Show me with your hand or by drawing.**

**What do you look for when you are trying to identify the greatest change? What about no change?**

Recap by asking students to review the vocabulary they used to describe graphs of change over time (e.g., "increase," "decrease," "no change") and specific amounts and times.

Recap the skills students used:

- Described the trends on a line graph
- Made both true and false statements about a line graph
- Analyzed false statements
- Sketched a graph given the story
- Plotted points to see a story

Students make a list of what they practiced in *Reflections* (*Student Book*, p. 147).

## Practice

*National Park Visitors,* p. 80
For practice in graph reading and analyzing statements.

*Salary History,* p. 81
For practice graphing a line and then answering questions about the story it tells.

*More Practice with Intervals,* p. 83
For practice changing scale on the $y$-axis.

## Extension

*Dropping Below Zero,* p. 85
For hints, graphs, and stories in which the $y$-axis includes negative numbers, and the line falls below the $y$-axis.

## Test Practice

*Test Practice,* p. 87

## Looking Closely

Observe whether students are able to

### Use new terms to describe graphs that show change over time

If students have trouble getting started, ask: "What do you see on the graph?" Whatever strikes them—the axes, the line, the labels—is an acceptable beginning point. Next tell students: "Say something true about what you notice. Now tell a lie about something you notice." Direct their attention to the vocabulary list from *Lesson 5* for ideas.

### Isolate and describe pivotal points in a graph of change over time

In discussion, ask questions about the graphs, such as

**Where is the greatest change?**

**Where is there no change?**

**What tells you that there was a sharp drop here?**

Students should be able to notice when there is a change, whether the change is an increase or a decrease, and whether it is significant.

### Refine ability to sketch line graphs based on oral information

At *Stations 1* and *2*, when students are telling each other stories and sketching them, look for a graph that you might want to share with the whole class at the end of the lesson. Check in with individual students to make sure they see how the direction and slope of the line indicate a change in trend. Ask the three questions above to assess students' understanding.

### Distinguish between true and false statements about a line graph

This activity relies on students' ability to analyze statements and provides an opportunity for you to learn much about who thinks critically, who is willing to accept a statement at face value, and who overthinks the statements. Your role may be to help students focus on the specific features of the graph, de-emphasizing background knowledge.

| WHAT TO LOOK FOR IN *LESSON 6* | WHO STANDS OUT? (LIST STUDENTS' INITIALS) | | | NOTES FOR NEXT STEPS |
|---|---|---|---|---|
| | STRONG | ADEQUATE | NEEDS WORK | |
| Concept Development<br>• Understands meaning of line, labels, and other features of the graph<br>• Sketches graph from story<br>• Explains change by comparing points or slope of line | | | | |
| Expressive Capacity<br>• Tells a clear story based on a graph<br>• Describes true and false statements about a graph<br>• Suggests corrections to make false statements true<br>• Expresses reasons to support finding a statement true or false | | | | |
| Use of Tools<br>• Uses rulers or other tools or strategies to ensure spacing is fairly equal | | | | |
| Background Knowledge<br>• Uses ideas and vocabulary from *Lesson 5* | | | | |

## Rationale

Precise descriptions of trends are important for interpreting graphs. As you work through this lesson, you and your students will use the language of line graphs to describe changes occurring over time. Students become familiar with phrases such as "steep increase," "steady increase," "slight increase," "sharp drop," "plunge," and "no change." The idea is for students to be able to read any graph of change over time and to make statements and inferences about it. By analyzing statements about graphs, students become more comfortable using information from them.

## Math Background

Some line graphs show change over a period of time, e.g., fluctuations in amusement park attendance, stock prices, or a baby's temperature. Labels give important information. The $x$-axis plots the time period; the $y$-axis depicts what is changing over time (e.g., attendance, price, temperature).

Sometimes bar graphs are used to represent change over time. If your students bring in examples of these bar graphs, you may want to use them to show how bar and line graphs can convey the same story.

Not all line graphs show change over time. In *Station 4: The Price of a Subway Ride*, the line graph shows a trend. When the price of a token goes up, it does not increase by pennies throughout a day of transition. There is an immediate jump in the graph.

Scale can be used to emphasize or stretch a story. Maintaining same-size intervals is assumed. On some graphs the axes do not start at zero but at other values, e.g., at "1940" or at "$2,000." This kind of graph is a close-up of a larger graph, and the scale continues from that point forward. Mathematicians use a squiggly mark when they want to vary scale within a graph. The graph to the right shows temperature with a break in scale to illustrate this point.

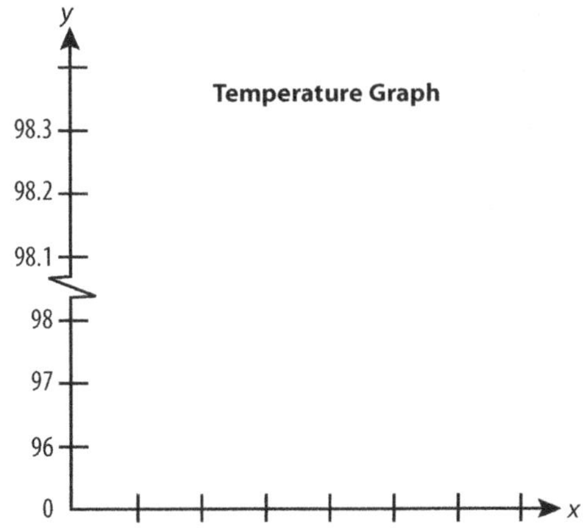


## Facilitation

If your class has difficulty working in pairs, try teams of three. While one student describes the story of a graph, a second student sketches a line describing the story, and a third student serves as an observer, paying attention to how the first two interact. Later, as part of the whole-class discussion, you can ask students how well they worked together and what ideas they have to improve communication. Another option would be to have one student describe the graph as the rest of the class including you, the teacher, sketches.

## Making the Lesson Easier

Guide students as a whole class through the activities in *Stations 1* and *2*. Students, in pairs, take turns describing and sketching line graphs. Then they practice plotting points and constructing intervals based on data.

## Making the Lesson Harder

Students can graph the data from *Jerri's Jobs* and *The Price of a Subway Ride* using Microsoft Excel™ or Google Sheets (see *Spreadsheet Guide, Student Book*, pp. 152–153).

*Notice that initially the assignment to write is confusing and intimidating to some students, as writing assignments in math class often are. Persisting, the teacher finds that many students do have statements to contribute. During the summary, students are able to respond to the question, "What is the point of this?" Observed by Martha Merson*

The teacher, Mary Jane, asks students to write three sentences about the graph.

Immediately there is tension at Sandra's table. Her frustration is palpable. She says aloud, "I don't understand," and repeats this many times during the next 10 minutes, in spite of efforts by others to help her get started. Another student says aloud, "I'm not sure why I'm doing this."

Ten minutes later, the teacher asks whether anyone has something to read. Irena reads her statement. Rhonda reads hers. Gina and Theresa, who did write and seemed to be on track, look confused and distressed, as if to say, "If this is what it was supposed to be, mine isn't. Do something."

Meanwhile, Edouardo says he has a statement to make, but says, "I didn't know how to write it down." He dictates his statement. At this point, Sandra herself says, "I'm starting to see." Mary Jane takes two more statements and then tells students, "Read the statements. Number your paper from 1 to 5. Write "T" or "F" next to each answer. Really be on your guard. This is like the GED, where you have to be precise about the words."

Before making public her own judgment, Mary Jane polls the class. The class is split on opinions. Mary Jane focuses attention on the first statement and calls on the people who thought the statement was false to suggest corrections. Rhonda changes the numbers to make one statement accurate. The students are quick to jump in with corrections. Time flies by.

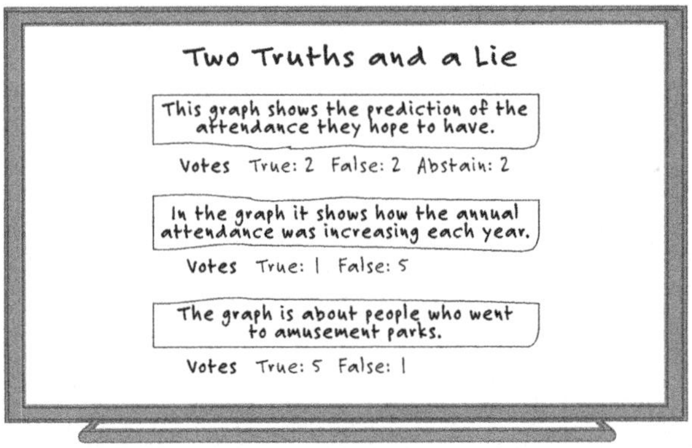


To summarize the class, Mary Jane asks, "So what's the point of all this?"

Irena answers cryptically, "Statistics," but Gina says, "Be careful what you read. Pay attention." Edouardo offers, "Look for what's missing," and Elvis says, "Look before you talk."

*Mary Jane Schmitt, Guest Teacher*
*Lowell Adult Education Center, Lowell, MA*

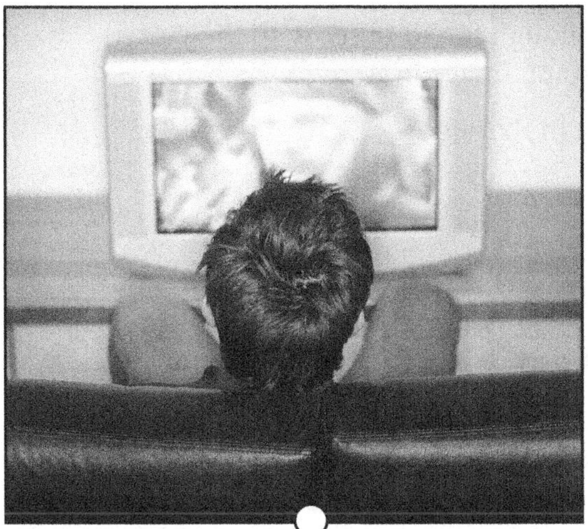

FACILITATING LESSON 7

# A Mean Idea

***How many hours on average do you watch TV?***

## Synopsis

The mean, one way to measure the center in a data set, is the focus of this lesson. Beginning with this lesson and continuing in *Lesson 9*, students describe numerical data sets with statistics.

1. The group agrees on a working definition for the terms **mean** and **average**.
2. Students predict and then calculate the mean; they use manipulatives to level, balance, combine, and redistribute, or otherwise explain their methods.
3. In pairs, students work with a data set to predict and calculate the mean, and experiment with adding one more value to see how the mean is affected.

## Objectives

- Use at least two strategies to determine the mean
- Define the word "average" as mean

## Materials/Prep

- Colored markers
- Easel paper
- Sticky notes
- Graph paper
- Rulers
- Counters such as pennies, plastic chips, or virtual sticky notes
- A graph with $x$- and $y$-axes large enough for all to see with the $y$-axis numbered 0–10 (see sticky-note graph image in *Activity 2*)

Collect student TV-viewing data for use in *Activity 1*.

## Opening Discussion

Open with the anecdote below, which then sets the stage for a discussion of TV-watching habits and the average number of hours per day most people in the United States watch TV. This story is also presented as a play that students can read aloud (*Student Book*, p. 89).

**Lola**: "I'm new to the U.S. and I live with and work for an American family. I am surprised at how much TV Americans watch! They watch from afternoon until bedtime. That is six hours a day for the children."

**Joanne**: "You can't base your idea of Americans on just one family. Is it possible they watch TV more on some days and less on others? In summer maybe the children watch less TV. One study I read said that the average American watches only three hours of TV a day."

**Lola:** "But an average is not a real person. I've seen real people in their own home."

Use this anecdote as the starting point for eliciting students' prior knowledge and ideas about averages.

**What point is Lola trying to make? How do people find averages?**

**Where have you heard the term "average"? What does it mean to you?**

**Have you heard the term "mean" used with a data set? Can you define "mean"?**

**How do people find averages?**

If students do not volunteer examples of their own, mention average temperature, average age, or average salary.

With students' input, develop a working definition for mean as an average representing the "center." Discuss how mean takes into account ups and downs over time, balances high and low numbers, or evens out amounts that are unequally distributed.

When people use the word "mean" in a mathematical context, they are indicating one kind of average. Some students might remember "mode" from *Lesson 1* or be familiar with "median." For this lesson, the words "mean" and "average" are interchangeable.

Begin the main activity by saying that data from the group on TV-watching will clarify the definition of mean.

## Activity 1: The TV Is On

Use the data about TV-viewing habits that students collected, or use the data outlined for Jess and Al.

| Jess | | Al | |
|---|---|---|---|
| Monday | 2 hours | Sunday | 10 hours |
| Tuesday | 2 hours | Monday | 5 hours |
| Wednesday | 4 hours | Tuesday | 4 hours |
| Thursday | 2 hours | Wednesday | 2 hours |
| Friday | 5 hours | Thursday | 4 hours |

Use sticky notes—one for each hour—to represent the number of hours watched each day, and arrange them on a graph like the one below in the starting demonstration example. Ask students how they could figure out the average daily TV-watching time for each person.

---

## Heads Up!

Take time to explore the numbers. Ask students to show their reasoning by demonstrating with pennies or sticky notes. Experimenting with numbers will be important in *Activity 2* when students identify a test score that fits and a data set that fits the median in *Lesson 9*.

---

If no one demonstrates re-allocating the sticky notes, model this with one set of data.

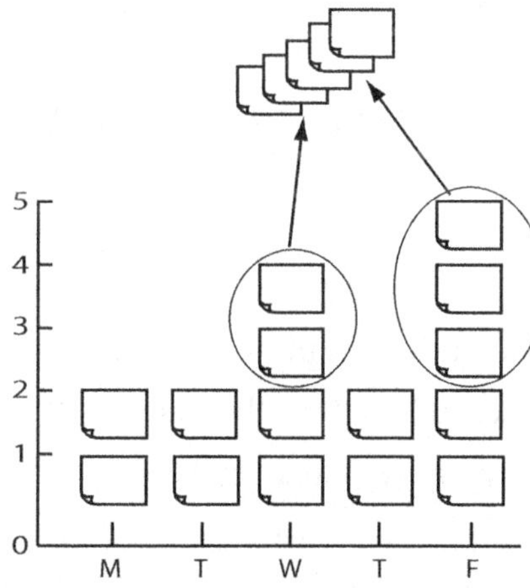


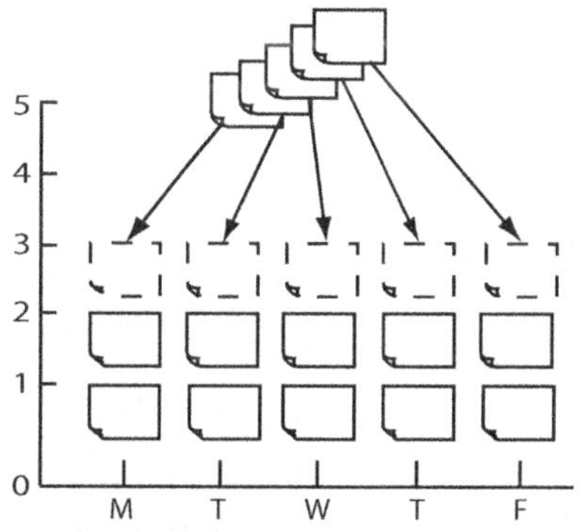


If students are not familiar with the algorithm, introduce it. Make the connection that counting all the sticky note hours, and redistributing them over five days is the same thing they are doing on their calculator when they total the number of hours and then divide by five. Then say,

**I need someone to show the data on a graph.**

After a volunteer plots the points on the large $x$- and $y$-axes, draw a dotted line to show where the average falls. Compare the average to individual data points.

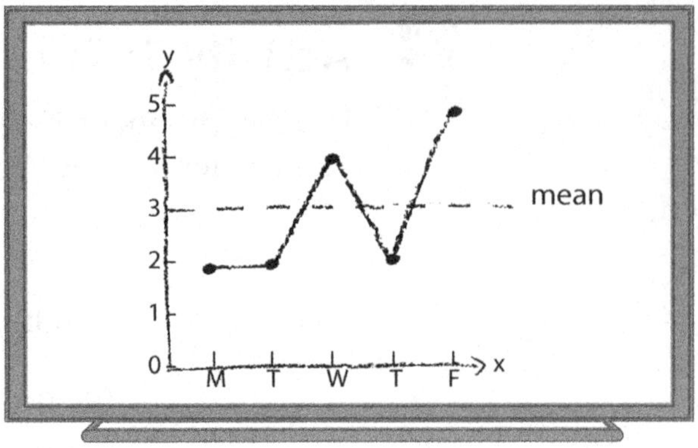


**How could you find the mean on a calculator?**

If students are not familiar with the algorithm, introduce it. First, count all hours on the sticky notes, and redistribute them over five days. Next do these same steps on the calculator by totaling the number of hours for all the days for each person and then dividing by five.

Before moving on, focus attention on each data set and the mean. Encourage students to describe the data set and to make generalizations. Model this by saying:

**In Jess's case, the mean is more than two (the smallest number of hours she spent in a day watching TV) and less than five (her highest number of hours in one day for watching TV). What about for Al?**

**Will the mean always fall between the largest and smallest values?** (Yes, unless all the values are equal.)

**In Jess's case, we took into account the five days of watching TV and divided all the hours among them. Will you always divide by five to find the mean?** (No, divide by the total number of items in the data set.)

## Activity 2: A Mean Score

This activity provides opportunities for students to find averages using calculators, reasoning, and graphs. Assign each pair of students one imaginary student's scores for fall (Polly, Malaika,Denise). Before they start working on Round 1: Fall Semester, direct each pair to

- Graph the scores;
- Predict an average;
- Draw a dotted line on the graph to predict where this average would fall;
- Find the average score based on the data; and
- Draw a line to show the actual average.

Students should check their own work or another group's work using a second method. Have students complete Round 2: Spring Semester after verifying their work from Round 1.

Two issues may arise as students work: (1) If students are unfamiliar with how to interpret decimals, you may need to explain how to round numbers. (2) A common mistake students make is finding the average of a set of scores, adding one more score to the existing average, totaling, and dividing by two. This will not yield an accurate average. If this occurs, ask:

**I want to see how a new score affects the average. What happens if I take the average of five tests, say 88, and then add a test score of 100, and divide by two?**

**Will you get the same answer as if you add all the scores, including 100, and then divide by six?**

**Why is there a difference between the two amounts? Which is the right average? Why?**

When pairs have completed at least one student's average for fall and spring, students report back. Ask them to describe the data set. For example: Denise took six tests. Her low score was 74, her high score was 91, and the mean was 85.

**What strategies did you use to arrive at the additional test score?**

**Why do you think your strategy worked? How does it relate to the rule for finding averages?**

Direct stuents to work on the additional questions in Rounds 3 and 4 following the same processes used in the previous rounds.

## Summary Discussion

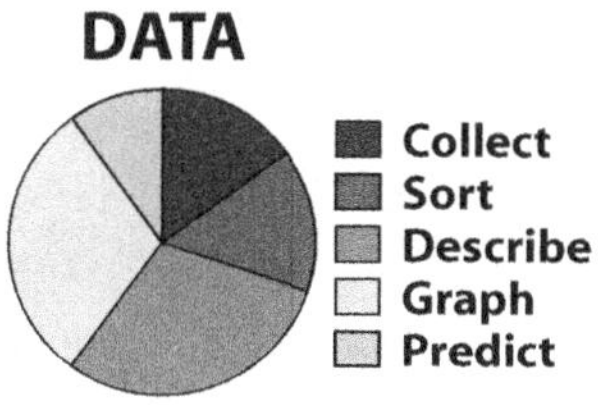


Using mean as the sole indicator of a data set could hide an **outlier**—an extremely high or low number—that distorts the impression of the data. Probe for examples by asking:

**What story does the mean tell?**

**Does the mean in any of these situations hide one data point that is dramatically different from the others?**

Refer students to *Reflections* (*Student Book*, p. 148).

## Practice

*Mental Mean,* p. 93
For practice finding the mean without the use of paper and pencil or calculator.

*Among Friends,* p. 94
For practice finding mean and using it for comparisons and to answer other questions.

*Data Collection,* p. 95
In preparation for *Lesson 8*, in which students collect high and low temperatures predicted for five consecutive days.

## Extension

*The Nurse's Mean,* p. 95
For practice using mean and using it to communicate a point.

## Test Practice

*Test Practice*, p. 97

## Looking Closely

Observe whether students are able to

### Use at least two strategies to determine the mean

Many adults know the algorithm for finding the mean: total of data divided by the number of data points. Seeing how this algorithm works with concrete objects may lead students to a deeper understanding of the divisor. Furthermore, they may begin to see alternative ways to check their work. The larger the numbers, the more useful it is to have an intuitive understanding. The better students' intuitive understanding, the more efficient they will be as test-takers and critical readers of statistical information. See *Practice: Mental Mean* (*Student Book*, p. 93).

Encourage flexibility by having students find the mean in different ways. Use the calculator to check the effect of one additional value or score on the mean. If students first make predictions, then use the calculator to compare the answers with their predictions. In time, the accuracy of their predictions should increase.

### Define the word "average" as mean

Do students recognize the terms "mean" and "average"?

Research by Brown and Silver (1989, p. 28) showed that "Most students in the 7th and 11th grades appeared not to understand technical statistical terms such as mean, median, mode, and range. However, there is evidence that they could compute the mean when asked for the average" (quoted in Zawojewski and Shaughnessy, p. 436). Use both terms—mean and average—in class to build students' familiarity. Mean is one kind of average. Mode, introduced in *Lesson 1*, can be reinforced any time a data set has a repeated value. Median, another type of average, will be introduced in *Lesson 9*.

| WHAT TO LOOK FOR IN *LESSON 7* | WHO STANDS OUT? (LIST STUDENTS' INITIALS) | | | NOTES FOR NEXT STEPS |
|---|---|---|---|---|
| | STRONG | ADEQUATE | NEEDS WORK | |
| Concept Development<br>• Finds the mean by computing<br>• Shows the mean by leveling<br>• Finds the mean mentally<br>• Makes a reasonable prediction for the mean<br>• Shows the mean in a line graph | | | | |
| Expressive Capacity<br>• Articulates how to find the mean<br>• Compares mean to individual data points | | | | |
| Use of Tools<br>• Uses pennies or Post-it Notes to show the mean by leveling<br>• Uses the calculator to find or check the mean | | | | |
| Background Knowledge<br>• Uses previous experience with mean to deepen understanding | | | | |

## Rationale

Average is a handy mathematical statistic. It functions as a shortcut—a quick way to describe a data set. Instead of listing all the data collected, researchers report an average. The mean gives the parent, consumer, or worker a quick way to picture the size of the average infant, an idea of the price of a gallon of gas, or the average time to commute to a job. Mean is a statistic that has a built-in flaw: Mathematically it may be the center, but in reality the number that is the mean may not appear anywhere in the data set. For example, the mean time to commute may be 25 minutes, but in reality, the trip takes 40 minutes at rush hour and 10 minutes at other times. The commuting time is never 25 minutes. Being able to recognize how the mean can give and yet disguise important information gives students another tool for critical interpretation of information and decision-making.

## Math Background

Finding averages can be conceived of as leveling. Suppose that the mean number of daily hours of TV viewing is five hours, and you have data for five days for one individual's TV-watching habits. It is possible that the person watched five hours each day ($5 + 5 + 5 + 5 + 5 = 25$; $25 \div 5 = 5$). It is also possible that the person watched 3, 4, 5, 6, and 7 hours on each of five days ($3 + 4 + 5 + 6 + 7 = 25$; $25 \div 5 = 5$); or 1, 3, 5, 7, and 9 hours on each of five days ($1 + 3 + 5 + 7 + 9 = 25$; $25 \div 5 = 5$); or 1, 4, 3, 8, and 9 ($1 + 4 + 3 + 8 + 9 = 25$; $25 \div 5 = 5$). There are many, many possible data sets, all yielding a *mean* of five. Notice that in the last data set, five is not even one of the data points, yet it is the mean.

The mean can distort people's impression of the data set when one or two data points are vastly higher or lower than the majority. These values are called outliers and, depending on the research question, they may be discarded during analysis. In some of these instances, the better measure for center is the median, which is the focus of *Lesson 9*.

Although mean may be used with no indication of any extreme data points in a data set, a discriminating interpreter should look for such information. *Activity 2*, along with the *Summary Discussion*, should highlight how an extreme data point will affect the mean.

## Context

For more information about TV watching and health, goals, and trends, see https://health.gov/healthypeople and https://health.gov/our-work/nutrition-physical-activity/presidents-council for the President's Council on Physical Fitness and Sports.

## Facilitation

Some students will only want to find the mean by adding the data points and dividing by the number of data points. While that is a correct method, stress that you also want them to find the mean in other ways, e.g., by mentally leveling, as in *Activity 2*, (see also *Practice: Mental Mean*, *Student Book*, p. 93). You want to make the concept of "measure of center" very transparent.

### Making the Lesson Easier

For those students who have difficulty finding the mean because of arithmetic issues, keep the numbers small. You might want to begin with the preceding example of TV-viewing hours: 3, 4, 5, 6, and 7. Encourage the use of concrete objects. You can show this with counters or sticky notes, or by adding and dividing by five.

### Making the Lesson Harder

Use numbers with fractions or decimals.

Take more time to discuss outliers. Should an outlier be eliminated from the data set? What are the repercussions of doing so? For example, if blood pressure readings for a person were all normal and then one were extremely high, what would you do? The average would not be dangerously high, nor would be the majority of the readings. However, a high reading might be indicative of a problem. There are no right, wrong, or easy answers to such questions. Each situation and each research question merits its own analysis.

Relate this discussion back to the comment in the story: In situations with an outlier, the mean is likely to be a number that does not appear in the data set. This is one reason why looking for a description of the minimum and maximum value is important.

During the debriefing, write the steps students said they took to figure out the mean. Use notation such as $(91 + 90 + 95 + 78 + 82)/5$. Use $X$ to indicate the score to be determined, for example: $(91 + 90 + 95 + 78 + 82 + X)/6 = 86$.

## LESSON 7 IN ACTION

*One student finds an application for mean. Observed by Martha Merson*

Sheila writes busily before class. She shows me a local newspaper article about 10 homicides. Police are looking for a connection. Sheila has made a table. It has the names of the victims, their ages, and causes of death. She also has a map of locations. She has figured out the mean and median ages of the victims. There is one age, five, that is an outlier. I say, "I wonder what would happen if you left that out." We do the math, and the mean jumps from 32 years old to 37 years old.

*Barbara Goodridge*
*Lowell Adult Education Center, Lowell, MA*

## FACILITATING LESSON 8

# Mystery Cities

***What about this graph is different from graphs you have seen in class?***

## Synopsis

Students examine climate graphs with two different but related types of information: precipitation and temperature.

1. As a class, students discuss mean temperature arrived at by finding the average high and low temperatures for the week.
2. Together they examine sample climate graphs featuring two types of data with changes in scale.
3. Students read graphs, match them to written descriptions of the climates represented in the graphs, and guess the cities described.
4. Then students check their answers with partners and discuss their reasoning.

## Objectives

- Read and interpret two sets of data on one graph
- Match text describing trends over time with graphs of the same phenomena
- Compare data across graphs of different scales

## Materials/Prep

- Calculators
- Masking tape
- Map of the United States
- Easel paper
- Recent weather data: five consecutive days of low and high temperatures plotted on a graph
- Weather information for students to react to in the Opening Discussion

## Opening Discussion

Display local weather information showing record highs and lows for today's date, as well as predictions for the day, night, and a week ahead. Ask students what they notice.

Ask:

**What is the difference between *climate* and weather?**

**What is *precipitation*? Do you think we have a lot of precipitation where we live?**

**Where is the climate vastly different from the climate here?**

Invite students who chose to collect non-local data in *Lesson 7*'s Data Collection (*Student Book*, p. 95) to show the location they choose on a map and share their data. Choose at least three locations to compare. Ask students::

**Describe how the weather would be different from the weather where we live.**

Invite students to give examples from other areas and articulate their reasons.

**Where did you find this data? Who else might be interested in knowing about climate in these different places? Why?**

If students cannot answer the first question, provide examples. Then invite them to give other examples and articulate their reasons. (See *Lesson Commentary*, p. 108, for ideas.)

Post the line graph of high and low temperatures that you prepared.

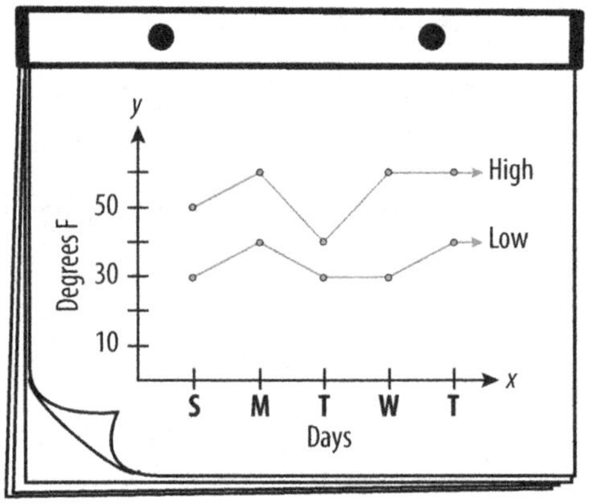

**What do you predict the mean temperature for the first day will be?**

**How can we find the mean?**

Review the steps, as necessary, and let students use calculators or another strategy for finding the mean. On the same graph, plot the mean temperature for each day and connect the points with a dotted line. Ask:

**What do you notice?**

If students do not mention it, point out where the mean falls relative to the other points. It should never be higher or lower than the highest or lowest points.

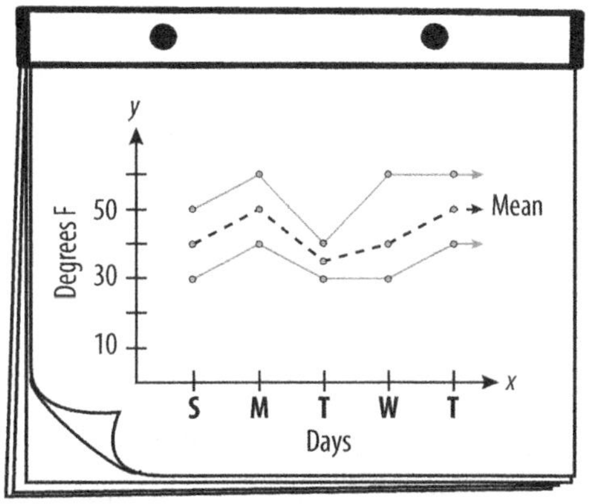


Students can practice plotting points and drawing the mean, using the data they collected on predictions for high and low temperatures.

Note that many weather stations collect more than two temperatures each day. Ask students:

**This is one way to find the mean temperature for one day. How would you find the mean temperature for an entire month?**

Take a few ideas and then move on.

## Activity 1: Reading Climate Graphs

Tell students that one way scientists describe the average weather in an area is with a **climate graph**—a graph showing the average temperature and average precipitation over 12 months.

Look at "Sample Climate Graph 1" (*Student Book*, p. 100) together, and ask students to describe what they notice about the way the graph is organized.

### Heads Up!

"Sample Climate Graph 1" shows precipitation (in inches) on the bar graph and temperature (in °F) on the line graph. For example, in July the average temperature was about 59°F with rainfall of less than 1 inch.

Now ask students to look at "Sample Climate Graph 2" (*Student Book*, p. 100), and ask:

**How are the two graphs the same?**

**How are they different?**

**Do you think the climate is the same in both locations? Why or why not?**

Ask students to record the temperatures and precipitation levels for both locations in February.

---

## Heads Up!

Note that scales in the two graphs are the same for temperature, and the shapes of the line graphs are quite similar. Then compare scales for precipitation and note the *big* difference between the two graphs.

---

Elicit the following observations about the graph sample:

- There are two kinds of numbers (and/or scales).
- There is a bar graph and a line graph.
- The letters on the bottom refer to months.
- Each bar represents a single average for that month.
- The line graph represents change of temperature from one month to the next. The bar graph represents change in precipitation from one month to the next.
- Both the line and bar graphs represent change over time.

## Activity 2: Mystery Cities

Present the activity by saying that descriptions and graphs for five cities became separated from each other, and the files with the names of the cities are missing completely. Students match climate graphs and weather descriptions and make guesses about what the cities are. Point out that descriptions of weather are subjective. Encourage students to be flexible as they consider how hot "hot" might be.

Allow students about 10 to 15 minutes to complete this activity individually, before checking their answers with partners.

Announce that you have found the files with the names of the cities. Post the names (see the list in the Heads Up! box on the following page) without revealing which city goes with which description and ask students to pair up, if they like, to figure out which cities match the graphs and descriptions.

Meanwhile, display the correct answers somewhere students can check them against their work when they are ready.

## Heads Up!

The matches are

Story of City 1 matches with Graph 4 – Phoenix, AZ

Story of City 2 matches with Graph 3 – Miami, FL

Story of City 3 matches with Graph 1 – San Diego, CA

Story of City 4 matches with Graph 2 – Boston, MA

Story of City 5 matches with Graph 5 – Seattle, WA

## Summary Discussion

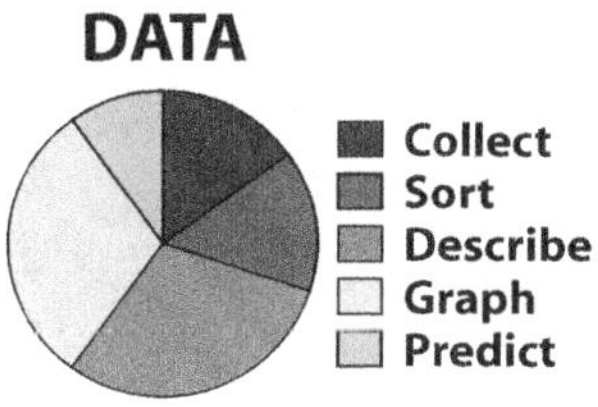


Debrief by asking students to explain their process:

**How challenging was it for you to read two sets of information on one graph?**

**What helped you read both sets of information together?**

**How did scale assist or get in the way in reading the graphs?**

**What worked for you in coming up with a match?**

Check in with individuals, particularly if they have a number of mismatches. Talk about what might have caused the errors and tell students that they will have a chance to keep working with some of this data again in the next lesson.

Refer students to *Reflections* (*Student Book*, p. 149).

## Practice

*Incomplete Climate Graphs!* p. 104
For practice completing lines, bars, and tables on unfinished graphs.

## Extension

*Quality of Life 1* and *Quality of Life 2,* pp. 106–107
For practice creating a graph with two kinds of information

## Test Practice

*Test Practice,* p. 109

## Looking Closely

Observe whether students are able to

### Read and interpret two sets of data on one graph

Can students compare two different data sets pictured on the same graph? Can they track the ups and downs of both sets of data using words and with the line and bar graphs?

If they seem to have difficulty with the two sets of data, encourage them to trace, with their finger or a colored marker, first one set of data and then the other.

### Match text describing trends over time with graphs of the same phenomena

Where do students seem to start: with the graph or the narrative? If students appear to be stuck, help them find a next step that continues from their instinctive approach rather than from your own.

Notice strategies successful students use and point these out to the whole group. For example, "I see Keesha taking notes. I see Maria sketching. I see John highlighting the names of the months and their seasons with one color." Ask questions to elicit peer learning: "Jose, how did you start?"

Check to see whether students are looking at both the precipitation and the temperature data for each climate graph as they match cities with the climate descriptions. Ask: "What are the characteristics listed? How would you see that on a graph?"

### Compare data across graphs of different scales

The task in this lesson is complex, and students may have difficulty with it. To handle the two sets of information, they must look at both the big picture and the details, specifically at the labels and scales. Recap the steps they have been taking when reading a new graph: looking at the title, forming an impression of the data, checking for labels, and looking at the scale and legend. If necessary, model by beginning to tell the story yourself, picking out some key points at the beginning, middle, and end.

| WHAT TO LOOK FOR IN *LESSON 8* | WHO STANDS OUT? (LIST STUDENTS' INITIALS) | | | NOTES FOR NEXT STEPS |
|---|---|---|---|---|
| | STRONG | ADEQUATE | NEEDS WORK | |
| Concept Development<br>• Reads labels on graphs<br>• Infers amounts from text descriptions<br>• Demonstrates flexible thinking (i.e., when moving between graphs is able to change expectations)<br>• Picks out the extremes on the graphs and descriptions | | | | |
| Expressive Capacity<br>• Tells a story using points on the graph, labels, trends, etc.<br>• Paraphrases descriptions | | | | |
| Use of Tools<br>• Observes changes in scale | | | | |
| Background Knowledge<br>• Draws on knowledge of maps, geography, and weather patterns | | | | |

## Rationale

In news stories and reports of all kinds, line graphs often present more than one trend or make more than one comparison. This enables readers to compare two similar but related events over the same time period.

This lesson helps students continue to develop fluency in interpreting and reading line graphs, and deepens their understanding of the connection between narratives and line graphs.

## Math Background

Students are looking at graphs with more information than any graphs they have either made or looked at thus far in this class. Several issues demand attention:

1. The bar graphs—what they represent and how they are made: The average precipitation for a specific month is used to determine the height of the bar. This is an example of how bar graphs are sometimes used to show change over time.
2. The line graphs—used here to show the average temperature for each month. The points are connected, month to month, showing the change, just as in the graphs that students made in earlier lessons.
3. The scales—on the horizontal axis: months; on the vertical axis: equal increments showing inches of rain and degrees F.
4. Two vertical axes—had the two graphs used like units (for example, both sets of data measured in inches), there would have been no need to have *two* vertical lines.

The exact values of the precipitation and of the temperature are not critical in these graphs. Approximations will do.

## Context

The U.S. Weather Bureau collects information on climate. To find the climate data, see newspapers, a library, or on the internet: http://www.wunderground.com or http://www.weather.com.

## Facilitation

Some students may be overwhelmed by the information in the climate graphs. The graphs are complicated—they have bars, lines, and labels on both the horizontal and the two vertical axes. Orient students:

1. Reassure them that it takes time to figure out graphs.
2. Focus students' attention on one part at a time. For example: "Look at the temperatures first. Where would you find them? What is the temperature in July? In February?"

3. Once they have figured out one part of the graph, move on to the second part.
4. Put it all together to see the bigger picture. For example: "In the summer months, the temperature was hot, and there was almost no rain."

Slow reveal graphs are a great way to help students identify key information in graphs. Find out how to use these with interesting current events at https://www.nytimes.com/2021/07/28/learning/introduction-to-whats-going-on-in-this-graph.html

## Making the Lesson Easier

Spend more time on the *Opening Discussion* and on *Activity 1*, making sure that everyone can handle the graphs. Pair students so they can read the descriptions together and discuss what they mean before trying to match the descriptions to the graphs. Provide extra copies of the descriptions and encourage students to cut out individual descriptions and match them with the graphs.

## Making the Lesson Harder

For those students familiar with spreadsheets, suggest they take the temperature data and class attendance and use a spreadsheet to make a graph similar to the climate graphs in this lesson—with two vertical axes. Be sure to explore how this works before asking students to create graphs with two vertical axes.

Brainstorm with students about other applications for graphs with two vertical axes, such as weather data and religious service attendance, or weather data and tips for restaurant workers.

Offer students the two *Extensions, Student Book*, pp. 106–107.

## LESSON 8 IN ACTION

*Knowing that the climate graphs would be unfamiliar to students, the teachers took an active role.*

I think this was new to all of the students—I had never seen these kinds of graphs before either. We anticipated that students would have trouble interpreting the graphs. We did a little group discussion first, introducing vocabulary.

I projected a weather map, and each student also had a copy of the United States map that showed all kinds of cities, but no state names.

I asked about 15 questions about what state a city was in, what was the temp today, etc. I used a *USA Today* paper. I also gave them a chart with cities and temps listed.

Then we handed out graphs and mystery city descriptions. Students worked together and alone for 15 to 20 minutes. Margaret and I circulated and tried to point out important things to look for. Then we wrote names of cities on the board, and students worked together to match up the cities. It went very well. A good math and geography lesson combined.

*Barbara Tyndall and Margaret Giordano*
*Intermediate Unit 13/Career Link, Lancaster, PA*

# FACILITATING LESSON 9

# Median

*What represents the median height?*

## Synopsis

In *Lesson 7*, students were introduced to the mean, a statistic that describes the average of a data set. The "Mystery Cities" climate graphs were based on the mean. In this lesson, the focus is on another statistic that describes the center: the **median**.

1. The whole class creates a working definition of median.
2. Students line up to find their median height and describe the data set.
3. The whole class finds the median in data sets with an odd and even number of points.
4. In pairs, students invent data to fit a given median.
5. The whole class reviews the meaning of median.

## Objectives

- Find the median for a data set with an odd number of data points
- Find the median for a data set with an even number of data points
- Present a possible data set given the median

## Materials/Prep

- Colored markers
- Sticky notes

Find information about median home prices in your area. Post quotes using the word "median," such as

- "Don't cross over the median."
- "The median age of a woman having her first child has changed dramatically in the past century in the United States."
- "The median monthly rent for a studio apartment in New York City is $2,800."

## Opening Discussion

Refer to three examples of the use of the term "median." Ask volunteers to read the quotes and to describe what they think is meant by the term. Ask students to volunteer other uses of the term.

---

### Heads Up!

Post a working definition of median, but do not explain how to find the median. That will happen in *Activities 1* and *2.*

---

End by making the point that the median is a measure of center. Tell students they will learn how to find the median in the activities that follow.

## Activity 1: Line Up

---

### Heads Up!

You need at least six students. For smaller classess, refer to *Facilitation* (p. 118) for suggestions. If necessary, join the class to ensure the first data set has an odd number of people.

---

Ask students to see whether they know who represents the median height in the class. Take their guesses and then ask them to line up in a row by height.

When students have arranged themselves, ask them who represents the median height for the class. Then ask them to start returning to their seats by twos, one from each end of the line.

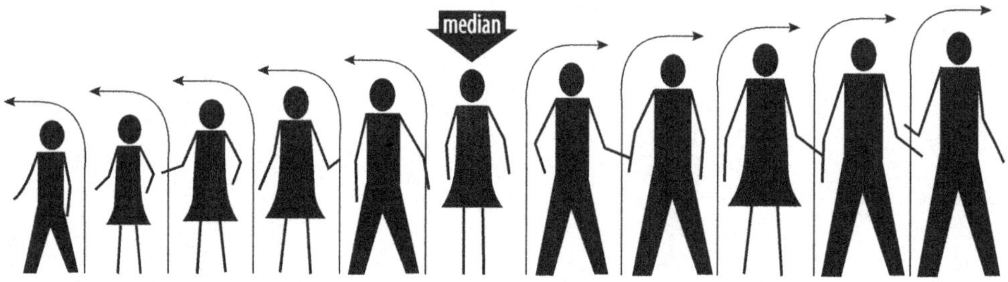

Because the number of participants is odd, there will be only one person left after the pairs take their seats. That student will represent the median height for the class. If a student of the same height as the one left standing has sat down, ask the class whether it would have made any difference if the two students had changed places (it would not, because the *height* is the median, not the student). Ask students to describe the data set, including the name of the person who represents the median height. See *Line Up* (*Student Book*, p. 112).

Find the median a few other ways. You could ask students to group by age (e.g., under 35 and 35+) and find the median height in each group. You could find the median height of the taller half of the class and the median height of the shorter half. You could use a different characteristic (e.g., median hair length). Students should be able to find the median with less help the second and third times.

Either add or eliminate one individual's height, and arrange heights once more, insuring that this time you have an even number in the data set. Ask students to predict:

**How will the median change if we add or eliminate one person?**

**Why do you think that?**

**How can you prove it?**

Even if students can answer the questions without lining up, it is a good way to check. In an even data set, there will be two students left, and the median should be halfway between those two. Ask students:

**How would the median change if the person we added were shorter or taller than the rest of us?**

Note that in a data set of heights, one new value will not affect the median dramatically and may not affect it at all. The impact of an additional value depends on the spread of the data points. For example, 1, 2, 3, 5, 7 is not a big range; 100, 37, 49, 72 is. A new data point can significantly impact the median in the second data set.

End with:

**Another way to find the median is to count the data points and figure out where the middle is located. That is the median, sometimes called the middle.**

## Activity 2: At the Party

Explain that the word "median" is used frequently without any information about the data set it came from. Being able to imagine a possible data set based on the median will make the statistic more meaningful.

Pair students. Assign one of the scenarios to each pair.

Explain the directions on *Student Book*, p. 113.

As students arrange the ages, observe critically:

- If there is more than one person of the same age, students may arrange the data points as in a frequency graph where x's are stacked. Each x still counts as one data point and can be removed from the stack starting at the top or the bottom.
- When students remove ends, make sure they are consistent in taking one from each end.
- If there are an even number of data points, and students end up with two sticky notes (ages), note whether they can determine that the median is halfway between those two ages. If the two ages are the same, the median is that age.

Questions to help pairs or to use during debriefing:

**How did the median help you think about the possible data set?**

**What was your first step in considering what the data set might look like?**

**How did you decide whether to use a data set with an even or an odd number of data points?**

**How could you check that the data set corresponds to your median?**

With the median ages established, ask:

**Suppose that my 106-year old aunt were added. Would your median age change?**

**Why?**

Note that the 106-year-old would not change the median age very much.

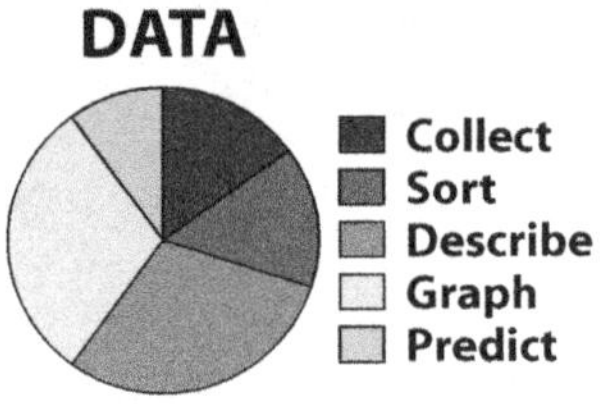


## Summary Discussion

Review how to find the median, given a set of data:

- Arrange the data points in order, either from smallest to largest or largest to smallest.
- Remove the ends, in pairs, until you have one or two data points left, or count the total number of data points and find the middle point in that number.
- If you were left with one item, that was the median. If there were two items, the median was the halfway point between them.

Review how to interpret a data set, given only the median. First ask:

**If I told you the median price for an apartment is $1,350, what would you know about the data set?**

Direct students to *Reflections* (*Student Book*, p. 149), where they will record their responses to the three main points about the median.

## Practice

*Finding the Median,* p. 114

## Extension:

*The Average Wage,* p. 115
For practice contrasting mean and median wages.

*Mean or Median?* p. 116

## Test Practice

*Test Practice,* p. 118

## Looking Closely

Observe whether students are able to

### Find the median for a data set with an odd number of data points

Since students often confuse median and mean, the two words are introduced separately at first. Do students grasp the idea of median as middle? Do they know how to find the middle when the number of data points is odd? Use a small data set at first and cross off pairs of endpoints until only one remains. Use sticky notes or counters such as pennies. Continue to use manipulatives until students are comfortable with the concept of crossing the numbers off.

### Find the median for a data set with an even number of data points

In an even data set there will not be *one* middle number. If the two numbers representing the median are the same, that amount is the median. If they are different numbers, the median is found by splitting the difference between them. In the number set 1, 3, 5, and 7, the two numbers in the middle are 3 and 5. Halfway between 3 and 5 is 4. Therefore, half the data items are less than 4 (1, 3) and half are greater than 4 (5, 7). It would not make sense to say that half the points are less than 3 (1), or less than 5 (1, 3).

### Present a possible data set given the median

Can students work backwards, using the idea of median, to create a varied data set? Or, do they choose numbers that are all equal to the given median (e.g., median = 26 and data set is 26, 26, 26, 26, 26); or many are the median, and a few are either above or below (e.g., median = 26 and the data set is 17, 26, 26, 26, 26, 98)?

| WHAT TO LOOK FOR IN *LESSON 9* | WHO STANDS OUT? (LIST STUDENTS' INITIALS) | | | NOTES FOR NEXT STEPS |
|---|---|---|---|---|
| | STRONG | ADEQUATE | NEEDS WORK | |
| Concept Development<br>• Finds the median in odd- and even-numbered data sets<br>• Invents data sets to yield a given median | | | | |
| Expressive Capacity<br>• Uses sound reasoning to explain median as the middle value of a data set<br>• Explains why median is an appropriate (or inappropriate) measure of the center of a data set | | | | |
| Use of Tools<br>• Uses Post-it Notes to demonstrate finding the median | | | | |
| Background Knowledge<br>• Brings working understanding of median from life experiences<br>• Uses frequency graphs to organize data before determining median | | | | |

## Rationale

Statistics of central tendency, like mean, are especially useful when communicating about a numerical data set. Median focuses attention on one middle value that can be easily communicated. Median is the preferred statistic for reporting average income and average home prices. A working familiarity with how median is derived makes the statements understandable.

## Math Background

People who report data have three options for communicating the idea of center for a data set: mode, mean and median. The median is different from the mean in that it is not affected by extreme values: the outliers. The mode represents the data point(s) that appears most frequently. The median might or might not be the data point that occurs most frequently. Although not all data sets have a mode, every numerical data set has a mean and median. The mode can be found for both numerical and categorical data (e.g., size: medium).

**Real Estate: Residential Sales**

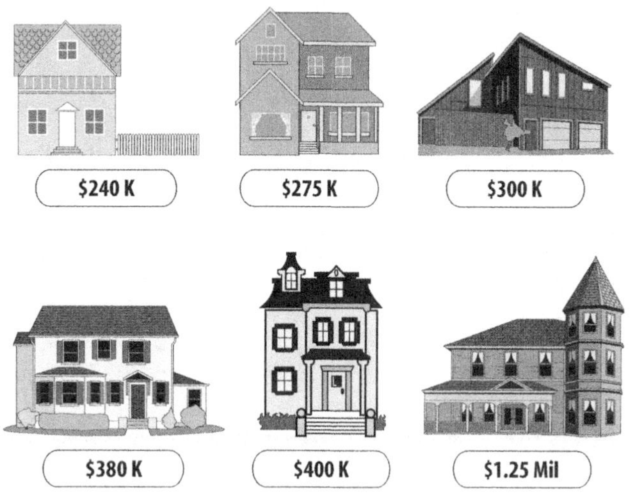


The median and the mean can both result in a number that is *not* part of the data set.

For example, when looking at housing prices in a given neighborhood, a data set might be $240,000; $275,000; $300,000; $380,000; $400,000; and $1,250,000. There are six data points (house prices), and the median is halfway between $300,000 and $380,000—that is, $340,000. Whether the most expensive house were $1 million or $10 million would make no difference. For that matter, if the least expensive house were $10,000, that would not affect the median either.

Now consider the mean for that data set. Add the numbers (house prices) and divide by six. Your answer is $474,166.70. That is certainly higher than the median, but, more importantly, it is not close to most of the data points, so it is misleading to say that it is the center. In this case, finding the median gives a better picture of the real cost of houses in that area.

# Facilitation

If students are few in number, provide an interesting data set of heights, e.g., 5′, 5′9″, 5′5″, 5′2″, 5′9″, 6′, and 5′6″. For further exploration, use heights listed in different units (all inches or all centimeters versus mixed feet and inches). Point out that working with one consistent unit will simplify the process. Subsequently, list heights with more variation, such as 4′11″, 6′2″, 5′1-½″, etc.

## Making the Lesson Harder

Students who quickly catch on to finding the median should try to identify both the mean and median and contrast them.

If students need help getting started, suggest lining up wages on a horizontal line or using one Post-it Note to represent each paycheck.

After students have finished, ask:

**Why do you suppose there is such a big difference between the mean and the median in the wages?**

**Which do you think best describes the wages? Why?**

**Which statistic would the owner choose to argue for not raising salaries? Why?**

*Students and teacher revisit the definition of average and create a set of salaries for a median of $30,000. Observed by Martha Merson.*

**Teacher**: "When has anyone talked about averages?"

**Student**: "Stocks. Dow Jones average. Teachers know average scores."

**Student**: "Median on the highway."

**Teacher**: "We should notice times when we see those words. I've seen them in references to real estate. Median price of a house. In the activity we have a median salary: $30,000. I'm going to show you one way *not* to do it. : We have to make up the salaries, and if all the salaries are $30,000, the median will be $30,000. How else could the workers get paid and still have $30,000 as the median salary?"

**Student**: "We can give however much we want?"

**Teacher**: "Yes, as long as the median is $30,000."

Students seem to know what is required, but there are several moments of tense struggle, not the typical buzz of productivity. Clearly students are not finding a way into the problem, and they are giving up.

**Teacher**: "Why is this difficult?"

**Student**: "Has to do with money. Not numbers."

The teacher provides a number line and positions the workers and their salaries along it, taking suggestions from the group.

**Teacher**: "Okay, what is the last salary?"

5K 27K 27K 27K 30K 30K 30K 30K **8K**

With one salary left, the students volunteer suggestions that will upset the sequence of the numbers and result in a lower median than $30,000. At first the teacher is reluctant to write in students' suggestions, but when she does, someone immediately sees the problem.

**Student**: "How about $50,000?"

5K 27K 27K 27K 30K 30K 30K 30K **50K**

**Teacher**: "Great. Do one more on your own."

*Barbara Goodridge*
*Lowell Adult Education Center, Lowell, MA*

# FACILITATING LESSON 10

# Stock Prices

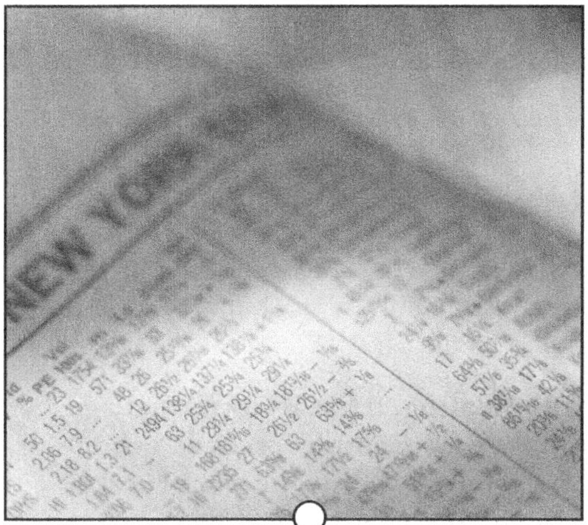

***What statistics are used to describe stock prices?***

## Synopsis

This lesson and the next give students opportunities to learn more about the stock market.

1. All students participate in a question-and answer-session to share knowledge.
2. Students examine a stock's performance first together and then independently by reading tables from a newspaper, isolating points on a graph, and filling in a table.
3. Students do three quick graphs with different scales, based on the same table.

## Objectives

- Use tables and graphs to understand complex information
- Apply an understanding of scale to compare information about different companies
- Use information from graphs to create tables

## Materials/Prep

- Calculators
- Local newspaper business section with stock market information
- Optional online sources such as https://www.NASDAQ.com; https://finance.yahoo.com; https://bigcharts.marketwatch.com; or a specific company's website under shareholder information (helpful for *Extension*).

Read *Hints for Teaching about Stocks* in the *Lesson 10 Commentary*, p. 128, before beginning.

## Opening Discussion

Start the discussion by acknowledging the mystique of the stock market and its effects on the U.S. and global economies. For example:

**For many of us, the world of stocks is intimidating. It has its own language, abbreviations, and mysterious fluctuations. Yet economic forces determine people's jobs, incomes, and quality of life. Understanding the stock market even a little demystifies some of the factors that affect our lives. People use information contained in charts and graphs to make investment decisions and predictions.**

Open with an informal question-and-answer session about **stocks**, **shares**, and the **stock market**. Listen for familiarity with the vocabulary. Ask students:

**What is the stock market?**

**What is a share of stock?**

**How does the stock market affect us?**

Listen to students' ideas; make sure they answer these questions simply. Refer to *Lesson 10 Commentary*, p. 128.

Together or in small groups, look at the table listing stocks included in the NASDAQ stock exchange (*Student Book*, p. 122). If you are working from a newspaper, make sure you are not looking at the mutual fund list. Mutual funds are combinations of several stocks. Note the table headings.

Model how to interpret one company's statistics. Ask students about the "–" and the "+" to ensure that everyone understands that these signs represent a decrease or increase in price, respectively.

For example, ABC Bop's stock price closed at $12.00. The low for the day was $11.19; the high was $12.10. The stock price was worth +0.51 (up $0.51) more than it was the day before.

### Heads Up!

If you have listings of stocks from local companies, they will provide immediate context for students, especially if any students are employed by any of the companies.

Call on volunteers or ask students in pairs to try interpreting a line of the table.

## Activity 1: Three Companies

Introduce the main activity. Tell students:

**Most people use graphs as a source of information about stock performance. We will look at three graphs to see the change in stock prices over the course of five days.**

Ask students:

**Do you think there is a great change or little change in the stock prices according to these graphs? Why?**

As students consider the amount of change, be sure they focus on the scale. Small intervals often influence how graphs are interpreted.

Discuss with students the need to pay close attention to scale so that they will not be misled. Paying attention to the labels on the $y$-axis and to the shape of the line is also important. Emphasize that a table provides supporting information to a graph, although it is not as helpful for getting an overall picture. A table has details about each of the data points, which can be used for making specific calculations.

Ask students to create tables from the graphs. Working in pairs, they should complete the table for each of the three stocks in the *Student Book*, or for another three that you or they choose.

---

### Heads Up!

Students might wonder how to isolate one point for their tables. Direct them to choose the point directly above the label for each day.

---

Students are asked to find the mean and median for three stocks. Regardless of scale, the mean and median can be found and compared. If they are similar, either one is a good measure for center. If they are very different, examine with students why that might be—is there one day with a much higher or lower price? If so, what happens to the mean and median if that day's price is considered an outlier and discarded from the data set?

## Activity 2: Altering Scale

This activity reinforces the impact of scale. Ask students to make three different graphs of the stock price for Energy Enterprises, each with a different scale.

Before students graph the information, ask for predictions:

**Which graph will have the flattest line? Which will have the steepest line?**

## Heads Up!

For the comparison to work, in every graph students must use each line on the grid paper. If, for example, they skip lines to make the graph with the interval of $25 more readable, the point will be lost. Therefore the *x–y* axes are given to students on page 125 (*Student Book*). The graphs should look something like this:

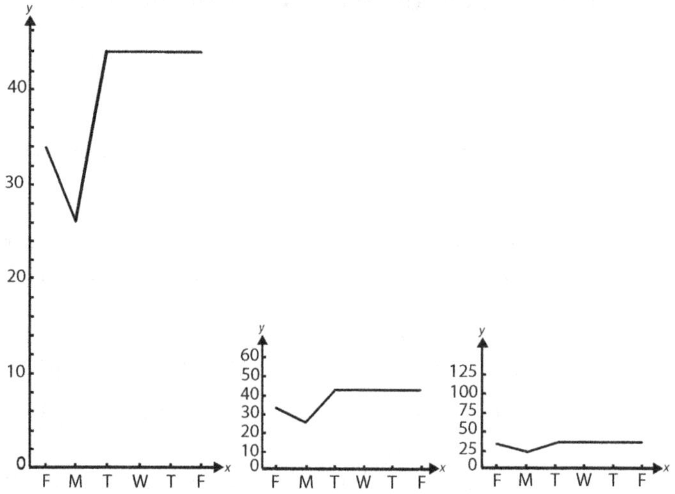


Ask students how the graphs differ:

**In general, what do you think is true about the scale?**

**In other words, the smaller the interval, the ...; the larger the interval, the ...**

**To give the impression of a flat line, I would want to make a scale that ...**

If students have a hard time articulating a generalization about the relationship between scale and the shape of the line, ask for a description of the graphs. Note that in each case, the line changed, but the data did not. Students will have another opportunity to experiment with altering scale to create an effect in the next lesson.

## Summary Discussion

Ask students to summarize what this lesson covered. They have

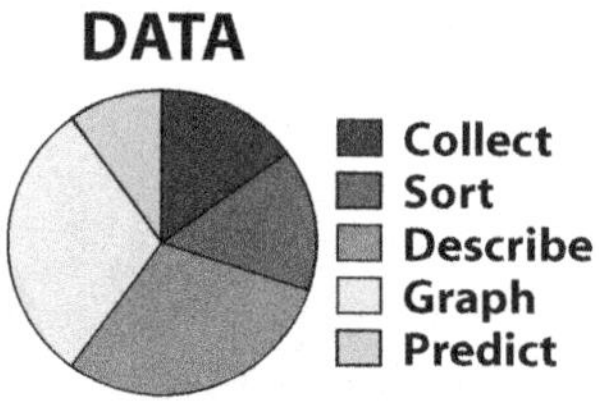


- Read tables with information on stock prices.
- Examined graphs of different companies over a six-day period, paying attention to scale.
- Made tables from graphs.
- Found mean and median stock prices and compared them.

Give students an opportunity to synthesize what they have learned. Ask:

**What are some companies that do business around here?**

**What do you think we could learn if we looked them up in the stock listings?**

**What do you think causes stock prices to rise or fall?**

Ask a few questions both to review the content of the lesson and to assess students' understanding. For example:

**What new information did you learn about the stock market?**

**What are you curious about?**

**Which is easier for you to use when reading about stocks, tables or graphs? Why?**

Students can record their responses in *Reflections* (*Student Book*, p. 150).

## Practice

*More Stocks—From Graphs to Tables,* p. 126

*From Graphs to Stories—Stock Prices,* p. 128

## Extension

*What Is New with the Stock Market?* p. 130
For practice doing research and making notes.

## Test Practice

*Test Practice,* p. 131

## Looking Closely

Observe whether students are able to

### Use tables and graphs to understand complex information

Can students easily interpret a table? Do some students need help tracking? Provide rulers or colored pieces of paper as a straight edge to help students line up.

Note whether students use monetary units (dollars and cents) or other ways of quantifying relative differences or changes as they analyze the tables.

Display a sentence stem to help students make statements using information across one row of a table.

## Apply an understanding of scale to compare information about different companies

Make sure that students pay attention to the scale on the vertical $y$-axis for the three different stocks. You should hear them use phrases such as "increased," "stayed the same," and "decreased." If they do not mention specific numbers, ask them by how much the stock increased or decreased.

Do students have strategies for comparing graphs with different scales? Suggest they put both lines on one graph. In trying to decide which line to add to which existing graph, a student has to think about the intervals. Will there be room on the paper for the scale the student has devised? It is likely that it will be easier to put the data with the smaller scale (intervals of every \$2, for example) onto the one with the larger scale (intervals of every \$10, for example). Another option is to extend the graph with the smaller scale by taping on additional paper.
One student invented this strategy so she could see \$500 on a graph numbered in \$1 increments.

The key is in the numerical details. Push students to look closely at the figures in their tables. For example, a \$2 change over five days may mean more than a \$12 change over the same number of days, depending on the price of the stock. This comparison calls for proportional reasoning.

## Use information from graphs to create tables

Students have had some experience isolating data points and describing what they mean. In this lesson, they learned a more systematic approach. Every line has an infinite number of points, each of which could be represented by an entry in a table. Picking out the relevant point or points is key.

Show students on the $y$-axis how to use the interval space (space between two points) to find an intermediate value. For example, if the interval is 10, start at the halfway point, 5, and move either up or down. Ask students how it might look on the $x$-axis using a different interval.

| WHAT TO LOOK FOR IN *LESSON 10* | WHO STANDS OUT? (LIST STUDENTS' INITIALS) | | | NOTES FOR NEXT STEPS |
|---|---|---|---|---|
| | STRONG | ADEQUATE | NEEDS WORK | |
| Concept Development<br>• Isolates points on a graph<br>• Uses *y*-axis to figure out value for points on the graph<br>• Moves back and forth comfortably between table and graph | | | | |
| Expressive Capacity<br>• Articulates observations about the companies' share prices<br>• Comments on the role scale plays<br>• Compares and contrasts mean and median share prices | | | | |
| Use of Tools<br>• Selects information from a table<br>• Uses paper, pencil, ruler, or other tool to keep track of his or her place | | | | |
| Background Knowledge<br>• Contributes to the discussion of the stock market based on prior knowledge of stocks<br>• Uses mean and median based on work in previous lessons | | | | |

## Rationale

Understanding even a little about how the stock market works demystifies some of the seemingly random changes that affect unemployment rates, entry level salaries, and minimum wage. Much of the information people use to analyze the economy is contained in charts and graphs. As students use line graphs to find information about a stock at a specific point in time and record that information in a table, they become more fluent in handling information encoded in either format.

## Math Background

Scale is one of the most powerful tools for manipulating the presentation of data on a line graph. The smaller the intervals on the *y*-axis, the more striking the change will seem. Imagine a graph of stock prices with labels on the *y*-axis for every $1.00. A change in stock price of anything more than a dollar will send the line of the graph up or down. Even a change of $0.75 will be apparent. If the intervals were $50 on a graph of the same dimensions, the line would vary, but the changes for any increase or decrease under $50.00 would be less dramatic.

With experience, your class will gain familiarity with manipulating scale. Until then, playing with the lines—placing both on the same graph, first with the increments smaller in scale, then with the larger ones—demonstrates how changing the shape of the line can be deceptive in influencing a reader's impression of the data.

## Context

You may want to have on hand a quick reference book on investments, such as *Investing for Dummies*, by Eric Tyson (New York: Wiley Publishing, 1999).

Students may be inclined to see playing the stock market as a quick fix for their financial worries. Stress that buying individual stocks is riskier and many people lose money that way. Investing in mutual funds is a way that people lower their risk because their money is spread out over many, many different companies.

## FAQs About Teaching about Stocks

**Q:** *What do I do if students ask me questions I cannot answer?*

**A:** Record the question and look up the answer later. Better yet, record the question and encourage students to find the answer themselves. Or record the question and refer to an expert. Use your own and your colleagues' networks to find someone knowledgeable about the stock market and willing to answer questions. Perhaps this person could visit the class.

**Q:** *This topic seems so far removed from my students' experience. Does it make sense to talk about stocks with people struggling at and below the poverty line?*

**A:** Yes! The classes that piloted the curriculum all had students in this same economic situation who welcomed the information. Some students have always felt closed out of the financial world and are excited to be sharing information on this topic. Other students already have stocks or the option to purchase them through a pension plan at work. Giving them more information on the stock market empowers them to ask more questions and to believe in their ability to understand the answers.

**Q**: *What are stocks?*

**A**: When a company wants to raise money, it sells stocks. In essence, it invites individuals in the general public to purchase a part of it, to become co-owners. The shares indicate ownership. The more shares, the more stake in ownership.

**Q**: *What is the stock market?*

**A**: The stock market is the place where people buy and sell stocks. There are stock markets both in the United States and abroad. Two of the more well-known stock markets are the National Association of Security Dealers and Automated Quotation (NASDAQ), where new stocks are traded, and the New York Stock Exchange, where more traditional, older companies' stocks are traded. Anybody anywhere can trade through a broker. Day traders use the internet or an app to trade without a broker.

**Q**: *What makes the stock market go up and down?*

**A**: In one class, students thought that production was the sole cause of fluctuation. In fact, many things can make a stock gain or lose value, for example, the demand for the service or product the company sells; the cost of the raw materials to make it; and the cost and availability of labor—the people to make the product or deliver the service. Generally speaking, the cheaper the product is to make and the more it sells, the more money the company makes and the more profits it can divide among its shareholders. Apart from these factors, the stock market is affected by "investment climate," a term that refers to the prevailing mood. When people feel uncertain about the future because of big issues such as possible war, inflation, or losing their jobs, they tend to sell stocks they own or hold onto their money rather than risk investing in the stock market.

**Q**: *How do people make money in the stock market?*

**A**: If you want to invest, you buy stock in a company by purchasing shares. Usually there is a minimum dollar amount or number of shares. You then might get a yearly or quarterly dividend—your share of the company's profit. In general, people try to buy stock cheaply and then hope that over time, the company will grow, sell more products or services, and be profitable. At some point, they might sell their shares for much more than they paid for them. When people talk about "getting in on the ground floor," they are referring to buying stock from a company before it becomes well known, then holding on to it until the company has worldwide recognition and profitability. Many people who bought stock in companies such as McDonald's and Microsoft before they were world-renowned have made a fortune on their holdings.

**Q**: *How do people lose money in the stock market?*

**A**: When you buy stock, you hope that the company will be profitable over time, but this does not always happen. The company you chose might develop financial problems. For example, in the early 1900s, people who bought stock in railroad trains found that when cars became more popular, the train companies lost money. Many companies whose business is conducted through a website or app have a great idea that does not become profitable. If a company cannot make money, it may close without ever making a profit. Investors (shareholders) do not get their money back if the company goes broke, or "belly up," to use a stock market term.

**Q**: *What do I do if students want to invest?*

**A**: Students need more information than you can give them. They need to be aware of the potential tax consequences, broker fees, etc. They can visit websites such as https://www.investopedia.com/, that can provide them with additional information about investing. Students can also find simulated investment sites on the internet.

## Facilitation

### Making the Lesson Easier

One teacher felt it was helpful to use a student's fantasy business to explain how to solicit investors, sell stocks, and build a business from one small store to a chain. You may need to depart from the mathematics to make sure that everyone understands the concept before you move on.

### Making the Lesson Harder

Stock prices fluctuate daily. A $1 fluctuation for one company can be a tremendous change, whereas $1 for another company is hardly significant. If your students have difficulty with this idea, connect it to their own personal experiences. For example, you could talk about buying products on sale. A $1 savings on a $2 item is considered a bargain. That $1 savings on something originally priced at $100 does not have the same significance. You can explore this with your class: Find the percent change for two stocks by dividing the change in price by the preceding day's share price.

# LESSON 10 IN ACTION

*Contagious enthusiasm with practical suggestions...*

I have a passion for investing and speculating in the stock market, and my enthusiasm was met by the curiosity and interest that already existed among the students. The students demonstrated a very high level of interest by asking many questions during the lesson and by participating during the summary discussion. I could tell from the questions being asked that the students did not have any direct experience with the stock market, nor did they have very much knowledge, only a vague definition of a "share of stock."

It was a fabulous idea to replicate the NASDAQ stock listings in the *Student Book*, and I enhanced the idea by buying one copy of the *Wall Street Journal* for every two people in the class. I explained the three stock exchanges and the Dow Jones Industrial Average. Then I had each pair of students calculate the Dow Jones Industrial Average by looking up the last price of all 30 stocks that belong to the Dow Jones Industrial Average. I instructed the students to total the closing prices for one share of each of the 30 Dow stocks and divide by the divisor (approximately 0.14) to adjust the prices for stock dividends and stock splits. We also examined a graph of the Dow Jones Industrial Average (found in every *Wall Street Journal*) and held a class discussion about why the graph went straight down after September 11, 2001, and why it eventually recovered.

I also very much liked the idea of having students make up a table from a stock chart. I handed out rulers to the students so that they could break the intervals on the $y$-axis into smaller intervals and therefore be more precise with their stock price estimates. Several class members thanked me for producing a lesson on the stock market.

*Elliot Fink*
*Borough of Manhattan Community College, New York*

STAGE | BROADWAY | SCREEN

VARIETY

PRICE 25¢

VOL. XCVII. No. 3 — NEW YORK, WEDNESDAY, OCTOBER 30, 1929 — 88 PAGES

# WALL ST. LAYS AN EGG

Going Dumb Is Deadly to Hostess In Her Serious Dance Hall Profesh

DROP IN STOCKS ROPES SHOWMEN

Kidding Kissers in Talkers Burns Up Fans of Screen's Best Lovers

Talker Crashes Olympus

MERGERS HALTED

HOMELY WOMEN SCARCE; CAN'T EARN OVER $25

# Facilitating Closing the Unit: Stock Picks

***Where would you put your money?***

---

## Synopsis

This lesson is an opportunity for students to synthesize what they have learned about line graphs, trends, axes, and scale. If you wish to do all three activities, anticipate that it will take longer than most lessons.

1. Students work in pairs to contrast five-year data on two different companies.
2. Pairs present graphs to the rest of the class and make a recommendation for one stock over another.
3. Students vote for their favorite stock, using the form provided, and give feedback about the strengths of each presentation.
4. The whole class participates in a review session.
5. Students complete tasks on the final assessment.

## Objectives

- Evaluate information and make recommendations
- Make an oral presentation demonstrating an understanding of change over time and the usefulness of tables and graphs, text, and scale to analyze data
- Read and solve problems involving data and graphs

## Materials/Prep

- Colored markers
- Easel paper
- A podium, microphone, or easel to hold graphs might make the final presentations more theatrical and fun. Consider how you will display the graphs students create to support their presentations. Four companies' information is included in *Activity 1*'s student pages, but students will be more inspired if they choose companies they connect with. Using the NASDAQ site to track a company's 5-year history is a simple but multi-step process; steps from *Student Book*, p.134, are below. If students research companies of their choice, the lesson will probably take two or three hours.
- Go to http://www.nasdaq.com and from the menu at the top of the page, click Market Activity.
- Think of a company whose stock you would like to research.
- Search for that company by typing the name into the search bar.
- Click on the company name and stock symbol you want to research.
- Locate time frames for the company above the chart and select a five-year time frame.
- Search for a second company and select a five-year time frame.
- The stock symbols of both companies will be listed above the most recently searched company name. Switch back and forth between companies by clicking the symbols to compare five-year trends.

Here are some ideas for products, along with company names and symbols.

| | |
|---|---|
| chocolate, e.g., Hershey | hsy |
| sneakers, e.g., Nike | nke |
| makeup, e.g., Avon | avp |
| ketchup, e.g., Heinz | hnz |
| home products, e.g., Martha Stewart Living | mso |

Copy the final assessment.

## Opening Discussion

Review the key points from the last class, especially the steps for moving from graph to table and the role of scale in determining the value of a share. These concepts are relevant to the main activity here.

Then explain the task:

**You will be given data on different companies. It will be your job to present the company of your choice as the best investment. Use what you know about change over time, data from the companies' past performance, and scale to convince the class to trust your recommendation.**

**What do we mean by "best" investment? This concept is something we need to agree on as a class.**

Brainstorm criteria that would make one stock seem like the best investment. If you have a long list, as a class, *prioritize three criteria*. Examples of criteria are stability, growth potential, past positive performance, and core service integral to our way of life. This list will help students think about what to consider when looking at stock investments.

Before the pairs begin, look at one of the graphs together and explain the labels on the horizontal axis. The division of the year into fiscal quarters may be a new concept for students.

## Activity 1: Stock Picks

Give students at least 20 minutes to work on their presentations. This should be enough time if students use the graphs they are given. While they are working in pairs, make sure students are keeping the criteria for best investment in mind.

---

### Heads Up!

Warn students when they are 5 minutes before presentation time. Ask them to write some notes to guide their opening and closing statements.

---

Before hearing the first presentation, help students fill in their voting sheets *Activity 1: Stock Picks–Voting Sheet* (*Student Book*, p. 138). Display creating a voting sheet with the company names and the three criteria students prioritized from ther Opening Discussion in the three middle columns. Tell students to rank each company from 1 to 5 in the three areas (1 being the lowest rating and 5 being the highest). Then students tally the votes to see which stock the majority thinks would make the best investment.

Also ask students to think about questions to pose once each pair finishes its presentation. Ensure that their questions are relevant by referring them to the criteria, and have them practice asking questions.

After each student or pair of students presents information and a recommendation for one company, the other students ask questions and then complete their voting sheets.

After all the presentations, students review how well each company did, based on the criteria. Then students go through the list and come up with an overall rank for the companies. There are enough lines on the voting sheet for 12 companies.

Discuss the rankings:

**Which stocks received a top rating from you? Why?**

**Why do you think the class does or does not have consensus on the ranking?**

**What were some things that convinced you to vote the way you did?**

Ask students to reflect on the process of planning and doing a presentation:

**How did the criteria you set at the beginning guide you as you compared graphs and planned your presentation?**

**How did you feel about trying to convince others your stock was the best choice, based on information from graphs?**

**What would have made this task easier for you?**

## Activity 2: Review Session

Transition to this activity by asking students to make observations about their own learning:

**In order to complete *Activity 1*, did you use skills that you didn't have before we started this unit?**

Build in a review session before giving the final assessment so students can practice study skills. Say, for example:

**One way to prepare for a test is to go over in your mind everything you have learned. List all the lessons we have covered so far.**

**What do you remember learning how to do?**

After brainstorming, direct students to one or more of the sheets designed to guide their review. *Activity 2: Review Session* (*Student Book*, p. 139), asks students to revisit the practice sheets in past lessons. Subsequently, direct students to *Reflections* (*Student Book*, p. 151), and ask them to consider what they would put in a portfolio of their best work.

## Activity 3: Final Assessment

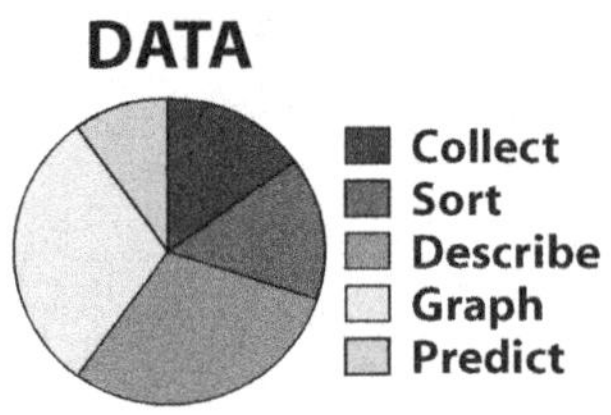


Ask students to check off the category that best describes how they feel about each problem: "Can do," "Don't know how," "Not sure" and then to complete the task if they can. When students finish, review their responses as a class.

## Summary Discussion

For closure, ask everyone to mention a favorite moment from this unit or to complete the sentence, "What I will remember about data and graphs is …"

## Looking Closely

Observe whether students are able to

### Evaluate information and make recommendations

By now students should be able to read graphs and think about what information they provide and the implications of that information. They should be able to see any particular patterns or trends in the data and whether there are any significant changes. They should also have learned to question what is happening when these changes occur.

The brainstorming activity done at the beginning of the lesson should indicate to you which criteria students think are important when looking at graphs related to stocks. If their suggestions for criteria seem inappropriate, discuss why others might be more appropriate.

### Make an oral presentation demonstrating understanding of change over time and the usefulness of tables and graphs, text, and scale to analyze data

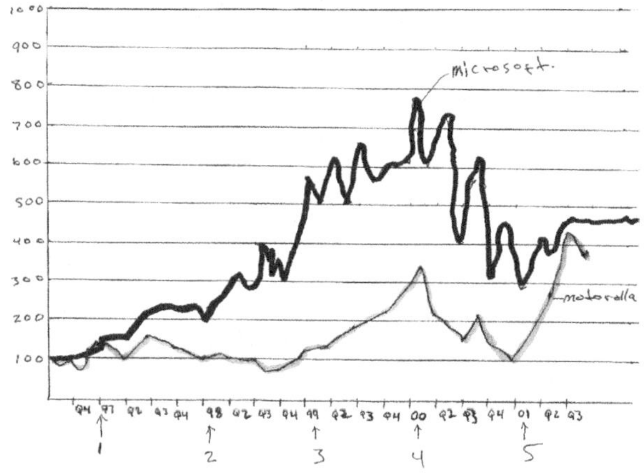


Students often have difficulty isolating the interval of greatest change. As they work, ask students to show you the period with the greatest change.

If they struggle, do a quick mini-lesson before students begin their presentations.

The student work pictured below shows that one pair who compared Hershey's and McDonald's stock gravitated to the bar graph format.

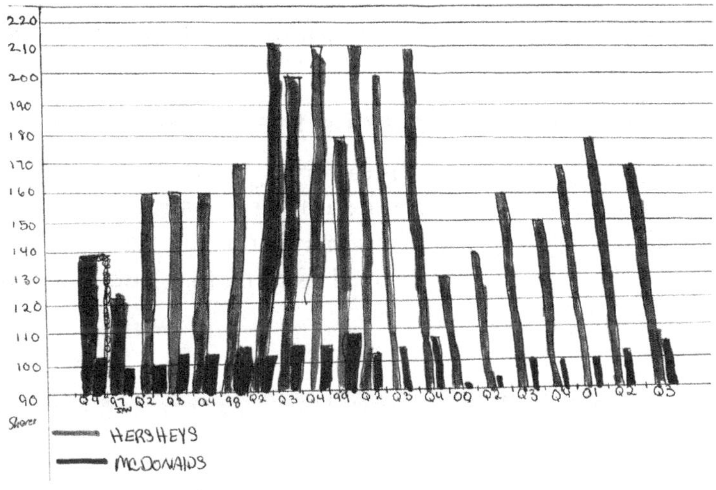


The bars highlight the difference in value between the two stocks, but with a bar graph it is harder to track the trend over time. A follow-up task for the students is to connect the bars with different colored markers so the audience can identify changes over time more easily. Often stocks' value increases or decreases in tandem due to world events. Ask students whether the stocks they compared did or did not show that and why. Stocks from the same market sector, such as consumer goods or technology stocks, will tend to fluctuate together. This is evident on both student graphs.

### Read and solve problems involving data and graphs

While students are completing the final assessment tasks, circulate, clarify tasks, and observe which ones cause students to struggle.

Check students' work against the *Final Assessment Checklist*, p. 164, to assess areas of strength and weakness. How do their confidence levels now compare to those in the *Initial Assessment*?

## Rationale

This activity is an application of learning from the past several lessons and requires relatively little new learning. It is a chance for you to observe how students integrate their learning. They can use each other as resources, but should rely on minimal help from you.

The presentations students make will include many of the ideas and tools students learned about and used in this unit. How these are put together will vary from one group to another, but several factors should be common to all. The presentation should have a clear goal; the data should make a direct point in a graph or table; the statements students write, using fractions or percents, should describe the points they are trying to make. Finally, the data must raise one new question.

## Facilitation

Because students work on their own or in pairs, you have an opportunity to observe how they tackle the task of comparing graphs.

Check in with each pair to make sure the students are comfortable with the idea of presenting in front of a group. You may need to offer some pointers to make them feel more at ease. Consider allowing students to record their presentation and play it for the class. See *Rubric for Presentations*, p.140.

### Making the Lesson Easier

It is easier to give students copies of the graphs than to find graphs online. If you think students will struggle with the ones provided in this lesson, you might try to find some that have a very similar (if not exact) scale. Then the issue of scale will be one less thing that students have to manage.

## CLOSING THE UNIT IN ACTION

*Students draw on their new knowledge and their prior experience during this rich discussion.*

For the most part, students were able to interpret the graphs and draw conclusions from them. In giving their reports, they drew on our opening discussion about what a potential investor needs to consider. They said things like, "Will people want the product in the future?" "Does the company have enough money?" "Did the company do well in the past?"

The students who compared Tommy and Nautica chose Tommy as the better buy, based on its superior past performance since the most recent price given for both was the same. Another student was quick to point out that you need to pay attention to other factors. Her teenage son and his friends, for example, no longer consider Tommy desirable, which could be a sign that the company will do worse in the future.

They had some difficulty drawing the two lines on one graph because they had trouble simplifying the lines (the graphs we used looked very much like those in the original lesson with weekly changes).

*Karen Johnsen*
*The Adult Learning Center College of Staten Island, New York*

## RUBRIC FOR PRESENTATIONS

### Students in the Group:

________________________________________

________________________________________

________________________________________

________________________________________

________________________________________

________________________________________

________________________________________

________________________________________

### Give 0–4 points for each area.

0 = not at all
1 = somewhat, but vague
2 = adequate
3 = well done
4 = exceeds expectations

___ Clearly stated what companies were being compared

___ Accurately described the trends of the stock prices over the past five years

___ Identified places on the graph where the greatest changes occurred

___ Made a recommendation

___ Used criteria to support reasoning

___ Used graphs to support the recommendation

___ Interpreted graphs correctly

### Extra

___ Designed graphs to supplement the presentation

___ Other

### Notes

________________________________________

________________________________________

________________________________________

________________________________________

________________________________________

________________________________________

________________________________________

Name ______________________ Date ____________

# Many Points Make a Point: Data and Graphs

## INITIAL ASSESSMENT

### Task 1: Decisions

1. Write down a decision that is coming up in your life, for example, which day care to place your child in; what brand of peanut butter to buy; when to take a trip to visit relatives; or where to buy a new sound system.

2. What criteria are important to you to consider in making your decision? Name at least two.

3. What information will you need to help you make your decision?

4. How will you organize the information?

## Tasks 2 and 3

For Tasks 2 and 3, look at each task, and decide whether you can do it, don't know how to do it, or are not sure. Check off the appropriate category, then complete the task, if you can.

___ Can do ___ Don't know how ___ Not sure

## Task 2: Printing Plant

Rosa is keeping track of mistakes at the printing plant. So far, here are the data she has collected.

**Week of January 17–January 21**

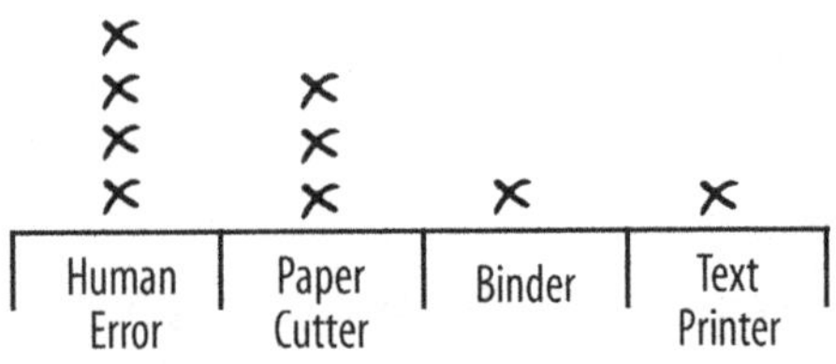


**Week of January 24–January 28**

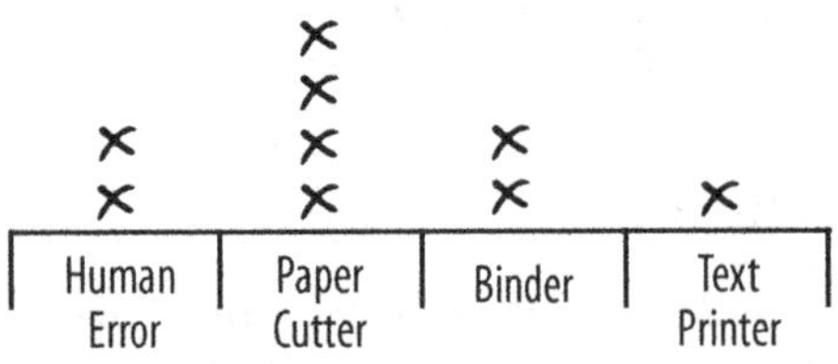


1. What do you notice about the mistakes?

   ___ Can do ___ Don't know how ___ Not sure

2. Regroup the data to show human error compared to machine errors. Show the data in three ways:

   a. Frequency graph

      ___ Can do ___ Don't know how ___ Not sure

   b. Bar graph

      ___ Can do ___ Don't know how ___ Not sure

   c. Circle graph

      ___ Can do ___ Don't know how ___ Not sure

3. Describe the data using fractions or percents.

   ___ Can do ___ Don't know how ___ Not sure

4. Choose one of your graphs and explain an advantage it has over the other kinds of graphs.

___ Can do ___ Don't know how ___ Not sure

5. Name one change Rosa could make to decrease the number of mistakes.

___ Can do ___ Don't know how ___ Not sure

## Task 3: Stocks, Tables, and Line Graphs

1. Write a sentence explaining what the graph shows.

___ Can do ___ Don't know how ___ Not sure

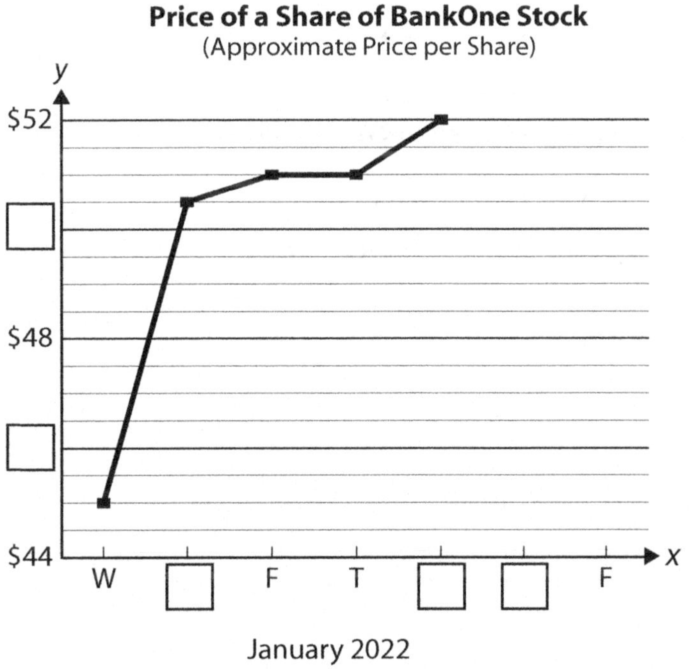

2. Make a table to show the price of a share of BankOne stock over five days in January 2004. (Saturday, Sunday, and Monday the stock market was closed.)

___ Can do ___ Don't know how ___ Not sure

| Day | Price of Share |
|---|---|
| | |
| | |
| | |
| | |
| | |

3. Add missing labels to the graph.

___ Can do ___ Don't know how ___ Not sure

4. Add the next two data points.

Thursday: $52.00

Friday: $51.50

___ Can do ___ Don't know how ___ Not sure

5. If you wanted to see the ups and downs more clearly, you would need a scale of at least __________.

___ Can do ___ Don't know how ___ Not sure

6. Determine the mean and the median for the data in the chart above.

a. Mean _______________ Median _______________

___ Can do ___ Don't know how ___ Not sure

b. The difference between the mean and the median is __________.

___ Can do ___ Don't know how ___ Not sure

c. Explain which most closely reflects the data.

___ Can do ___ Don't know how ___ Not sure

## CLASS TALLY

### Task 1:

Decisions Listed

| Task | Can Do | Don't Know How | Not Sure | Percent Who Can Do |
|---|---|---|---|---|
| 2. 1 | | | | |
| 2 | | | | |
| 2a | | | | |
| 2b | | | | |
| 2c | | | | |
| 3 | | | | |
| 4 | | | | |
| 5 | | | | |
| 3. 1 | | | | |
| 2 | | | | |
| 3 | | | | |
| 4 | | | | |
| 5 | | | | |
| 6a | | | | |
| 6b | | | | |
| 6c | | | | |

### Notes on Confidence Levels

## INITIAL ASSESSMENT CHECKLIST

Use a ✓, ✓+, or ✓– to assess how well students met each skill. When you give feedback to students, note areas in which they did well in addition to areas for improvement.

Use ✓ to show work that is mostly accurate; some details, additional work needed

Use ✓+ to show work that is accurate, complete.

Use ✓- to show work that is inaccurate, incomplete.

Student's Name______________________________

| Task | Skills | Lesson Taught |
|---|---|---|
| **1. Decisions** | | |
| ______ | Applies data and graphs to own life | *Reflections* |
| | Gathers appropriate information to inform decision | *Midpoint Assessment* |
| **2. Printing Plant** | | |
| 1 ____ | Explains main idea/story of graph | 1, 2, 5 |
| 2 ____ | Regroups data | 2, 4 |
| 2a ____ | Creates frequency graph | 1, 2 |
| 2b ____ | Creates bar graph | 3 |
| 2c ____ | Creates circle graph | 3, 4 |
| 3 ____ | Describes data using fractions and percent | 2, 4 |
| 4 ____ | Explains advantages of one format, compare and contrast | 3, 4 |
| 5 ____ | Interprets data to make a decision | 2 |
| **3. Stocks, Tables, and Line Graphs** | | |
| 1 ____ | Explains main idea/story of a graph | 5, 6 |
| 2 ____ | Makes a table from graph | 10 |
| 3 ____ | Adds missing labels | 5 |
| 4 ____ | Graphs the point(s) on a graph | 6 |
| 5 ____ | Scales or adjusts scale | 10 |
| 6a ____ | Finds mean and median | 7, 9, 10 |
| 6b ____ | Compares mean and median | |
| 6c ____ | Explains how mean and median relate to data set | |

## OVERALL NOTES

### Strengths

______________________________

______________________________

### Areas for Improvement

______________________________

______________________________

Name ______________________________ Date ______________

# Many Points Make a Point: Data and Graphs

## INTERIM ASSESSMENT

In this activity, you will work alone or with a partner to analyze data and present a decision you would make based on the data.

You may want to review your notes from previous lessons before you start. For example, do you remember three ways to organize data? How did you use fractions and percents to describe data?

### Decisions with Data

Choosing an apartment:

| | | |
|---|---|---|
| cost | location | facilities |
| reputation of landlord | other | |

Choose two criteria that are important to you. What information would you collect?

Choosing a new partner:

| | | |
|---|---|---|
| skills for earning income | attractiveness | kindness |
| reputation | age/experience | other |

Choose two criteria that are important to you. What information would you collect?

## Activity: The Data Say

Your group should follow these directions:

- Look at the story and data set.
- Think about how this information could be used based on the story provided.
- Create either a bar or circle graph using some of the data. You will *not* use all of the data when you create your graph. Pick the data that fit your needs.
- Look closely at the graph you have created. What does the graph suggest to you? How do the graph and the data help you address the issues?
- Prepare a presentation. Share your story, and present your graphs and your response to the issue. Provide reasoning for the format you chose.

When your group presents, tell the class:

- The story of your data.
- The topic of your graph.
- The time period of your data.
- What the information suggests.
- Your recommendations for a next step.
- Other information you wish you had (or questions the data raises).

Do not forget to do the following:

- Include everyone from your group in the presentation.
- Be clear about what the information in the graphs represents.
- Explain how you organized the data.

### Activity: Option 1, A Look at Your Program

As a student you are part of a larger adult education program and may have ideas for how to improve it. Looking at program data may support your ideas or lead to new ideas.

Work with your teacher to fill in the charts below with statistics from your program.

### Employment Status

| Full Time | Part Time | Not Working | Total |
|---|---|---|---|
| | | | |
| | | | |

| Caregiver for Child Under 18 | Non-Caregiver for Child Under 18 | Total |
|---|---|---|
| | | |
| | | |

| Types of Classes Offered by Program | Number of Students |
|---|---|
| | |
| | |
| | |
| | |
| | |
| | |
| | |
| | |

Make a graph using some of the data related to your ideas to improve the program.

Could this information be used as a new way to recruit students or to design a new class? Be creative.

What question related to the data do you have about the program?

## Activity: Option 2, Your Survey

Survey 10 people, asking each one the same two questions.

Some ideas:

How many years were you in school?

How did you get to school today?

Are you in favor of . . . ? (Choose a topic like school uniforms.)

Organize the data.

Some people understand information better when it is presented in pictures. Make a graph using some or all of the survey data.

Based on the data, write three statements that summarize what you have learned. Use fractions or percents in your statements.

## Activity: Option 3, Crime Fighter

Imagine that you live in a city of 41,000 people. You recently read an article in your local paper about crime. The police reported the following incidents over a period of two weeks:

- 12 stolen vehicles
- 11 residential break-and-enters (6 committed at night, 5 during the day) with 9 resulting in theft
- 5 store robberies (3 armed and 2 unarmed)
- 22 motor vehicular break-and-enters (15 committed at night, 7 during the day), but none resulting in theft from the vehicle
- 6 larcenies from vehicles (break-and-enters with theft)

As a concerned citizen of this city, you want to see less crime.

Organize the information into a graph. Think about how many categories you will use.

Then plan a meeting with the Crime Watch officer to talk about what you think could be done about crime in your city, using the data to make your case.

## Activity: Option 4, Eating Out

A new restaurant in town would like to advertise to adults. They asked 20 people who live in the area, "Where do you hear about new places to eat?" This is what the restaurant survey revealed:

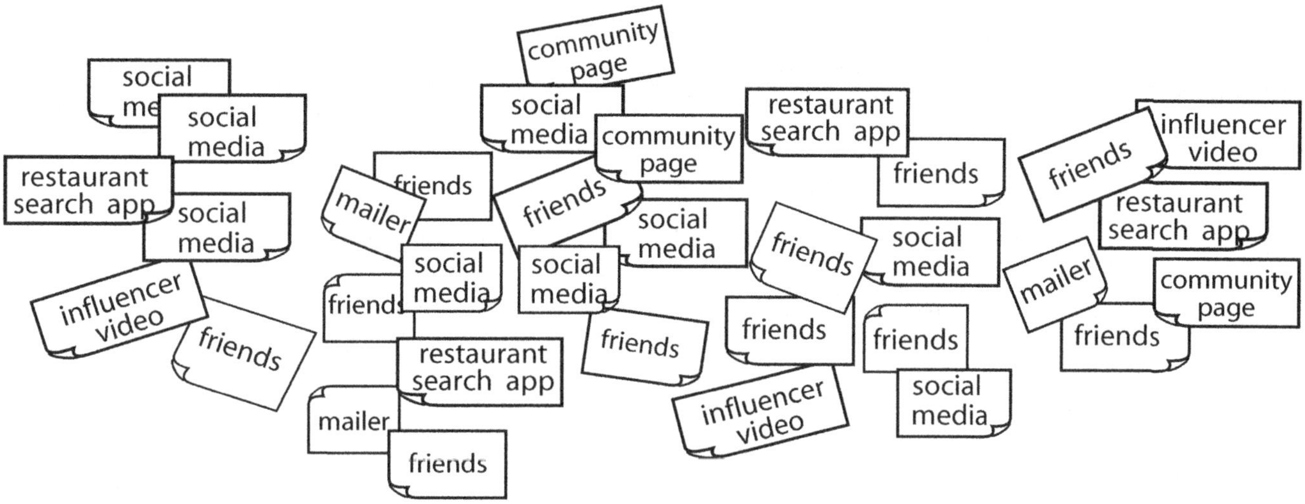


Create a bar or circle graph of the data.

Based on the responses, what suggestions would you give the restaurant owner?

Where do you recommend the restaurant owners put their advertising dollars? Should they advertise a special offer of any kind?

## Practice: Decisions, Decisions

Circle the top two criteria or types of information that will be meaningful for you in making a decision.

Then write down at least two places where you can find the information. Choose from the following: talking to others; looking at prices, stickers, and labels; calling an information line; checking with a reference librarian; using a printed resource; using an online resource; or reading the newspaper. You may also think of additional resources.

The first table presents an example.

### Buying a Car or Motorcycle

| Criteria | Where to find information |
|---|---|
| Price | Check stickers, phone dealers, and ask a friend who works on a car lot |
| Gas mileage | Check Internet, check magazines on cars for reviews of models I am interested in |
| Size | |
| Total mileage | |
| Repair record | |
| Warranty | |

### Choosing a School

| Criteria (Circle 2) | Where to find information |
|---|---|
| Reputation | |
| Location | |
| Languages/programs | |
| Facilities | |
| Faculty | |
| Financial aid available | |
| Other ________________ | |

### Choosing a Health Plan

| Criteria (Circle 2) | Where to find information |
|---|---|
| Cost | |
| Locations served | |
| Flexibility | |
| Other ________________ | |

### Choosing a Mortgage Company

| Criteria (Circle 2) | Where to find information |
|---|---|
| Cost | |
| Reputation | |
| Easy to understand materials | |
| Other ________________ | |

### Choosing a Bank

| **Criteria** (Circle 2) | **Where to find information** |
|---|---|
| Rates | |
| Location | |
| ATM convenience | |
| Easy to understand materials | |
| Staff/languages | |
| Reputation | |
| Other ____________________ | |

### Choosing a Lawyer

| **Criteria** (Circle 2) | **Where to find information** |
|---|---|
| Rates | |
| Reputation | |
| Location | |
| Experience with the type of case | |
| Helpful staff | |
| Languages spoken | |
| Age | |
| Other ____________________ | |

### Choosing a Microwave

| **Criteria** (Circle 2) | **Where to find information** |
|---|---|
| Cost | |
| Reputation | |
| Experience with brand | |
| Warranty | |
| Features/flexibility | |
| Other ____________________ | |

### Choosing the Timing for Having a Medical Procedure

| **Criteria** | **Where to find information** |
|---|---|

### Choosing a Babysitter

| **Criteria** | **Where to find information** |
|---|---|

Name ______________________________ Date ______________

# Many Points Make a Point: Data and Graphs

## FINAL ASSESSMENT

### Task 1: Decisions

1. Write down a decision that is coming up in your life, for example, which day care to place your child in; what brand of peanut butter to buy; when to take a trip to visit relatives; or where to buy a new sound system.

2. What criteria are important to you to consider in making your decision? Name at least two.

3. What information will you need to help you make your decision?

4. How will you organize the information?

### Task 2: Mind Map

List all the words and ideas that come to mind when you think of *data and graphs*.

**DATA and GRAPHS**

## TASKS 3–7

For Tasks 3–7, look at each task, and decide whether you can do it, do not know how to do it, or are not sure. Check off the appropriate category, and then complete the task, if you can.

___ Can do ___ Don't know how ___ Not sure

### Task 3: Snowfall Data

___ Can do ___ Don't know how ___ Not sure

Here is the snowfall for a town in New York state.

| | Snowfall in inches |
|---|---|
| Week 1 | 19 |
| Week 2 | 0 |
| Week 3 | 8 |
| Week 4 | 14 |

Make a graph to show the snowfall data.

## Task 4: Tables and Line Graphs

1. Write a sentence explaining what the graph shows.

    ___ Can do ___ Don't know how ___ Not sure

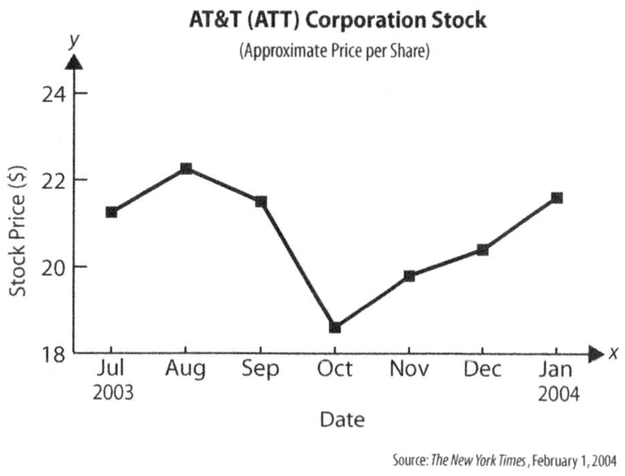


2. Make a table for the price of AT&T stock each month.

    ___ Can do ___ Don't know how ___ Not sure

| Month | Price of Share |
|---|---|
| | |
| | |
| | |
| | |

3. If you wanted to see the ups and downs more clearly, you would need a scale of __________.

    ___ Can do ___ Don't know how ___ Not sure

4. Name the ways you could report the average cost of a share.

    ___ Can do ___ Don't know how ___ Not sure

5. The following questions refer to the price of a share of BankOne stock.

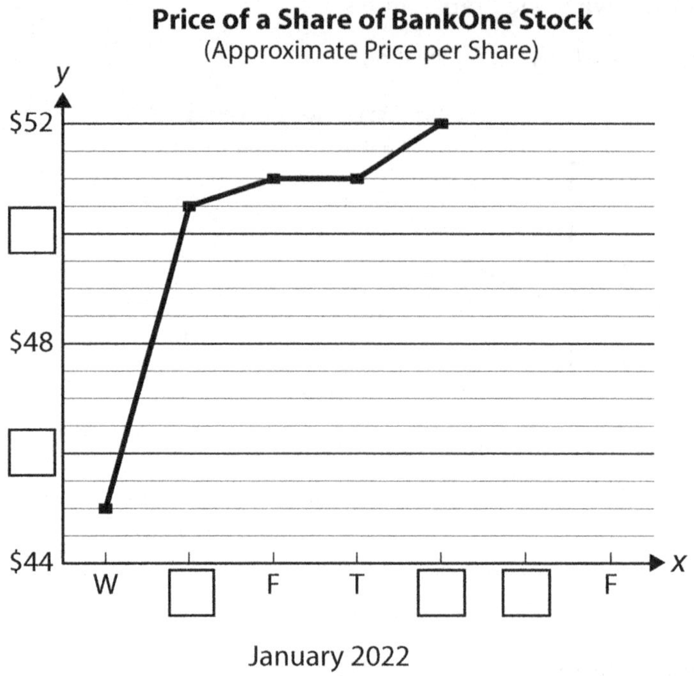


a. Determine the mean and median price of a share of BankOne stock during this period.

Mean _______ Median _______

___ Can do ___ Don't know how ___ Not sure

b. What is the difference between mean and median? _______

___ Can do ___ Don't know how ___ Not sure

c. Explain whether the mean or median seems to most closely reflect the data.

___ Can do ___ Don't know how ___ Not sure

## Task 5: Printing Plant

Rosa keeps track of the number of rejected books at the printing plant where she works. Here are the data she has collected so far.

| Date | Number of Rejects | Date | Number of Rejects |
|---|---|---|---|
| Monday, June 3 | 4 | Monday, June 17 | 2 |
| Tuesday, June 4 | 5 | Tuesday, June 18 | 3 |
| Wednesday, June 5 | 2 | Wednesday, June 19 | 1 |
| Thursday, June 6 | 3 | Thursday, June 20 | 3 |
| Friday, June 7 | 8 | Friday, June 21 | 10 |
| Monday, June 10 | 5 | Monday, June 24 | 4 |
| Tuesday, June 11 | 5 | Tuesday, June 25 | 5 |
| Wednesday, June 12 | 6 | Wednesday, June 26 | 5 |
| Thursday, June 13 | 7 | Thursday, June 27 | 2 |
| Friday, June 14 | 12 | Friday, June 28 | 14 |

1. What do you notice about the data?

   ___ Can do ___ Don't know how ___ Not sure

2. What is the mode?

   ___ Can do ___ Don't know how ___ Not sure

3. Comment on what the mode indicates.

   ___ Can do ___ Don't know how ___ Not sure

Rosa took a closer look at the rejects. The frequency graph shows the cause of rejections for the three days with the highest number of rejects.

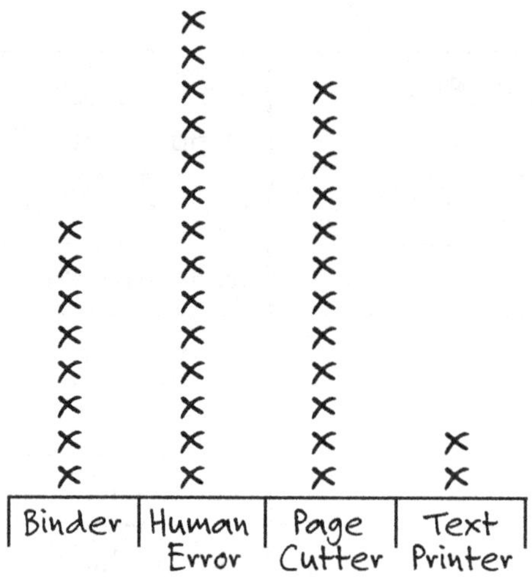


4. Analyze the data and then circle the change you think would make the biggest difference in the number of rejects.

   ___ Can do ___ Don't know how ___ Not sure

   (1) Training for workers

   (2) A new binder

   (3) A new text printer

   (4) A new page cutter

   (5) A new binder and text printer

5. Regroup the data into machine errors and human error. Show the data in three different ways:

   a. Frequency graph

      ___ Can do ___ Don't know how ___ Not sure

   b. Bar graph

      ___ Can do ___ Don't know how ___ Not sure

   c. Circle graph

      ___ Can do ___ Don't know how ___ Not sure

6. Describe the data shown in your graphs, using fractions or percents.

   ___ Can do ___ Don't know how ___ Not sure

7. Choose one of your graphs and explain an advantage it has over the other two graphs for presenting these data.

   ___ Can do ___ Don't know how ___ Not sure

## Task 6: Story Graph Match

Tony, an on-site trainer, is preparing for his annual review. He has several types of data and wants to present them to his boss.

1. Find the one story that best matches each graph. Write the letter for the story above the graph.

 ___ Can do ___ Don't know how ___ Not sure

_______ _______

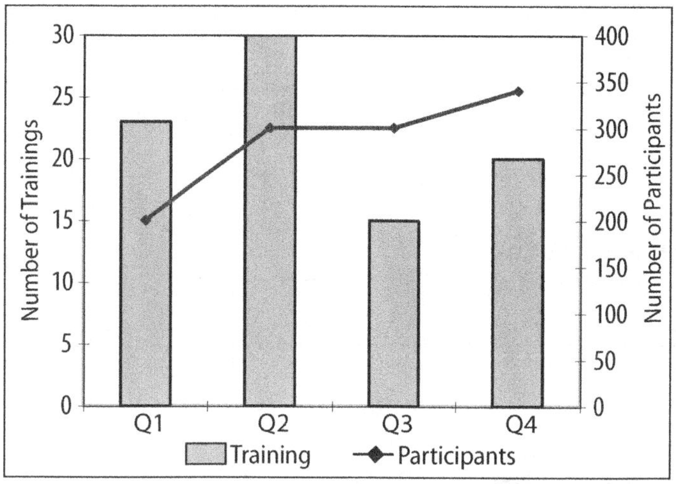


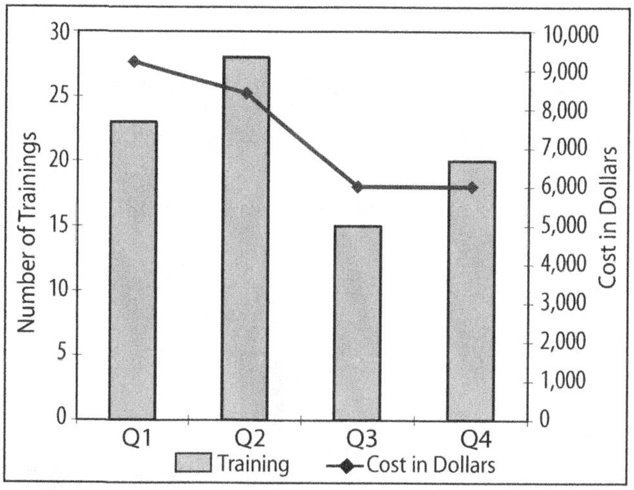


a. Cost increased while the number of trainings Tony provided decreased.

b. Tony cut costs and provided 15 or more trainings each quarter.

c. Tony's number of participants decreased as the number of trainings decreased.

d. In the fourth quarter, Tony kept the number of participants steady and cut the number of trainings in half.

2. Which graph and story make a stronger case for Tony's good work. Explain.

 ___ Can do ___ Don't know how ___ Not sure

## Task 7: Test Type Questions

Choose the correct answer for each of the problems below:

1. Which of the following statements can be made based on the graph?

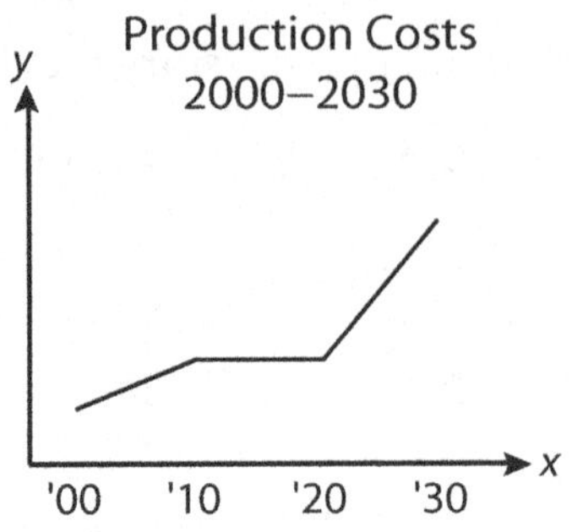


___ Can do ___ Don't know how ___ Not sure

(1) The cost of production decreased steadily between 2000 and 2030.

(2) Production increased steadily between 2000 and 2030.

(3) Production costs soared between 2010 and 2020.

(4) Production costs more than doubled between 2010 and 2020.

(5) Production soared between 2020 and 2030.

Problems 2 and 3 are based on this graph:

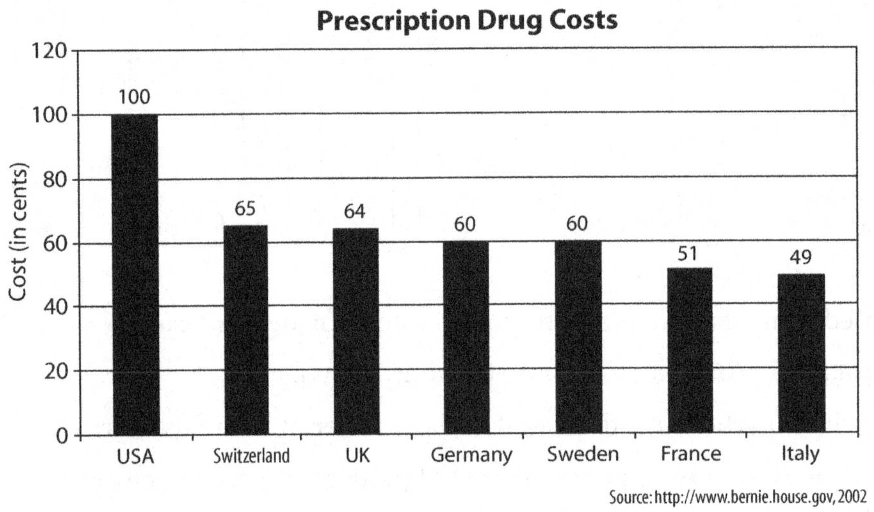


2. Which of the following countries averages about half of what prescription drugs cost in the United States?

___ Can do ___ Don't know how ___ Not sure

(1) Switzerland

(2) The United Kingdom

(3) Germany

(4) Sweden

(5) France

3. According to the graph, on average, for every dollar an American pays for prescription drugs, someone from Germany would pay about how much?

___ Can do ___ Don't know how ___ Not sure

(1) $600

(2) $60

(3) $6

(4) $0.60

(5) $0.06

Problem 4 is based on this graph:

**Budget for Catty Cove Corporation**

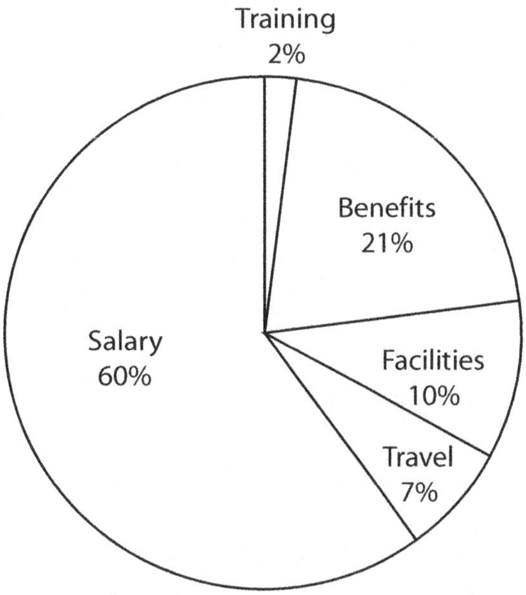


4. Which of the following represents a little less than one-quarter of the training budget for Catty Cove Corporation?

___ Can do ___ Don't know how ___ Not sure

(1) Salary

(2) Training and Benefits

(3) Facilities and Benefits

(4) Training

(5) Travel and Training

## FINAL ASSESSMENT CHECKLIST

Use a ✓, ✓+, or ✓– to assess how well students met each skill. When you give feedback to students, note areas in which they did well in addition to areas for improvement.

Use ✓ to show work that is mostly accurate; some details, additional work needed.

Use ✓+ to show work that is accurate, complete.

Use ✓- to show work that is inaccurate, incomplete.

Student's Name______________________________________________

| Task | Skills | Lesson Taught |
|---|---|---|
| **1. Decisions** ____________ | Applies data and graphs to own life<br>Gathers appropriate information to inform decision | *Reflections*<br>*Midpoint Assessment* |
| **2. Mind Map** ____________ | Makes a list of terms, e.g., types of graphs, purposes for data and graphs | *Vocabulary*<br>1, 2, 3 |
| **3. Snowfall Data** ____________ | Plots points using data from a table | 6, 9 |
| **4. Tables and Line Graphs** | | |
| 1 ____ | Explains the main idea/story of a graph | 5, 6 |
| 2 ____ | Makes a table | 10 |
| 3 ____ | Suggests a change in scale to change shape of line | 10 |
| 4 ____ | Lists types of averages | 7, 9 |
| 5a ____ | Finds mean and median | 7, 9 |
| 5b ____ | Compares mean and median | 10 |
| 5c ____ | Explains appropriateness of mean or median | 10 |
| | | 6 |
| **5. Printing Plant** | | |
| 1 ____ | Describes data in table | 10 tables |
| 2 ____ | Explains mode | 1 |
| 3 ____ | Comments on what mode indicates | 1 |
| 4 ____ | Knows what changes the data support | 1, 2 |
| 5a ____ | Shows regrouped data in frequency graph | 1, 2 |
| 5b ____ | Explains bar graph | 3 |
| 5c ____ | Explains circle graph | 3, 4 |
| 6 ____ | Describes data using fractions and percents | 2, 4 |
| 7 ____ | Chooses format and explains advantage | 3, 4 |

| | | |
|---|---|---|
| **6. Story Graph Match**<br>1 ____<br>2 ____ | Matches the story and graph<br>Makes an argument with graph and story | 8<br>*Closing the Unit* |
| **7. Test Type Questions**<br>1 ____<br>2 ____<br>3 ____<br>4 ____ | Explains main idea or story of a line graph<br>Reads $y$-axis to determine value of a point<br>Uses benchmark fractions and percents to describe data | 5, 6<br>3, 6<br>1, 2, 4 |

## OVERALL NOTES

### Strengths

____________________________________________________________

____________________________________________________________

____________________________________________________________

____________________________________________________________

____________________________________________________________

____________________________________________________________

____________________________________________________________

### Areas for Improvement

____________________________________________________________

____________________________________________________________

____________________________________________________________

____________________________________________________________

____________________________________________________________

____________________________________________________________

____________________________________________________________

____________________________________________________________

### Fatal Occupational Injuries 1990–2000 Logging vs. Grocery Store Work

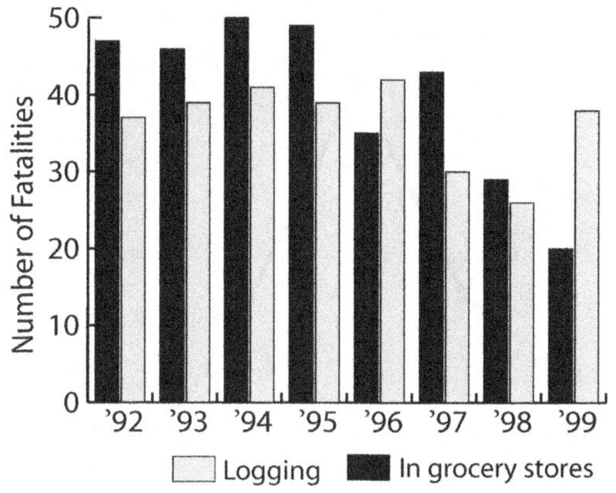


Source: U.S. DOL, Bureau of Labor Statistics Data, 2002

### Unemployment Rate: 2008–2018

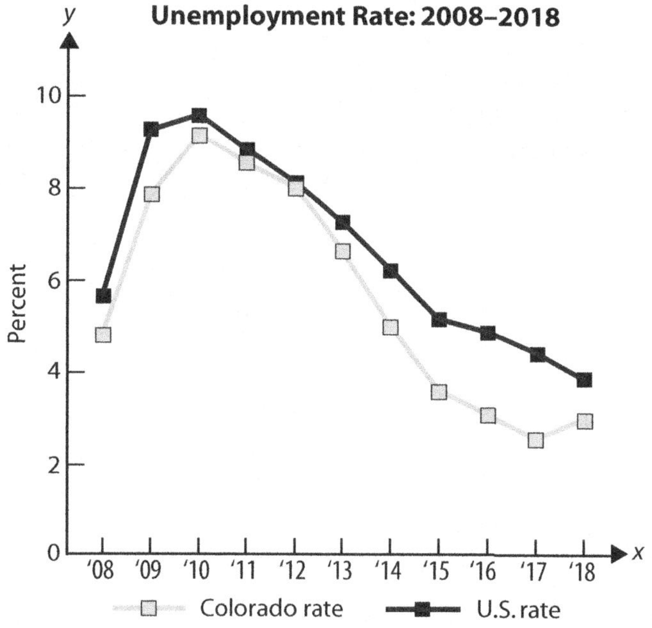


Source: U.S. DOL Bureau of Labor Statistics Data, 2022

### Smoking-Related Deaths

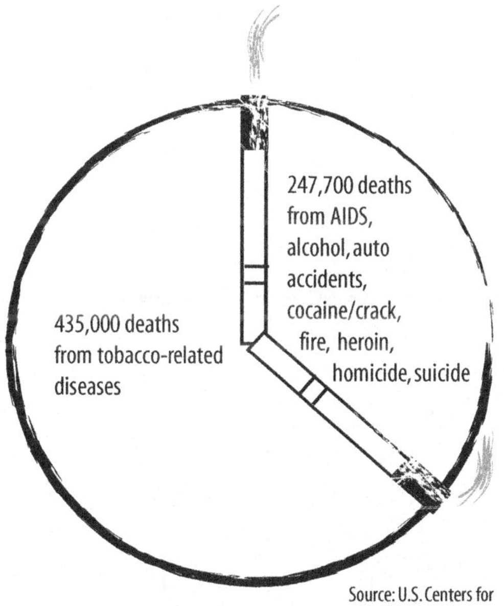


Source: U.S. Centers for Disease Control, 2000

More Americans die from smoking every year than drugs, AIDS, car accidents, alcohol, suicide, homicide, and fire combined.

### Donors' Blood Types

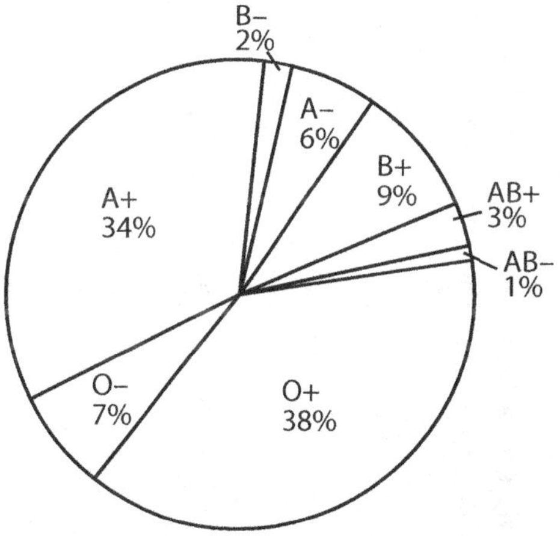


Source: American Association of Blood Banks, 2003

# Blackline Master 2

## The Changing Needs of the American Workforce

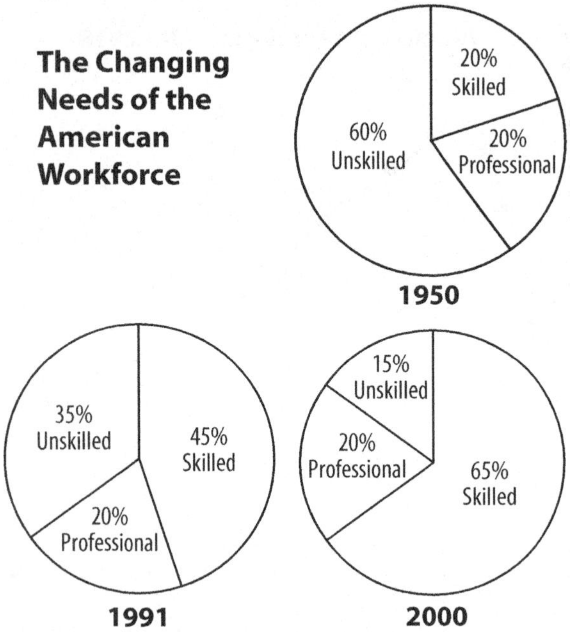


Source: Association for Career and Technical Education (www.acteonline.org)

## Immigration from China 1850–1900

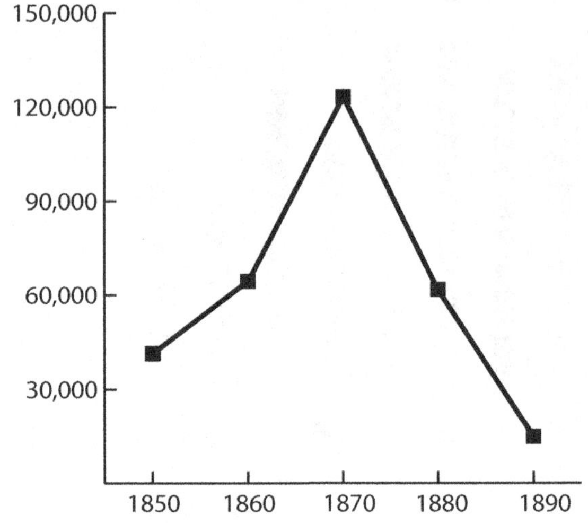


Source: Department of Homeland Security, 2003

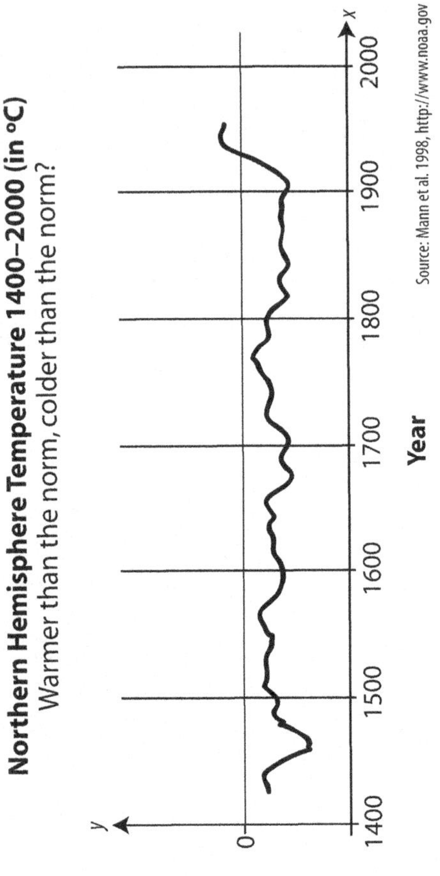


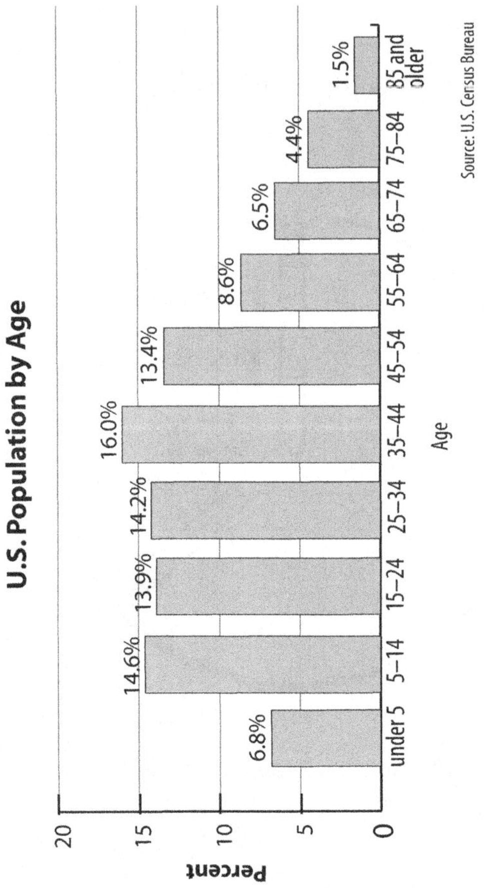

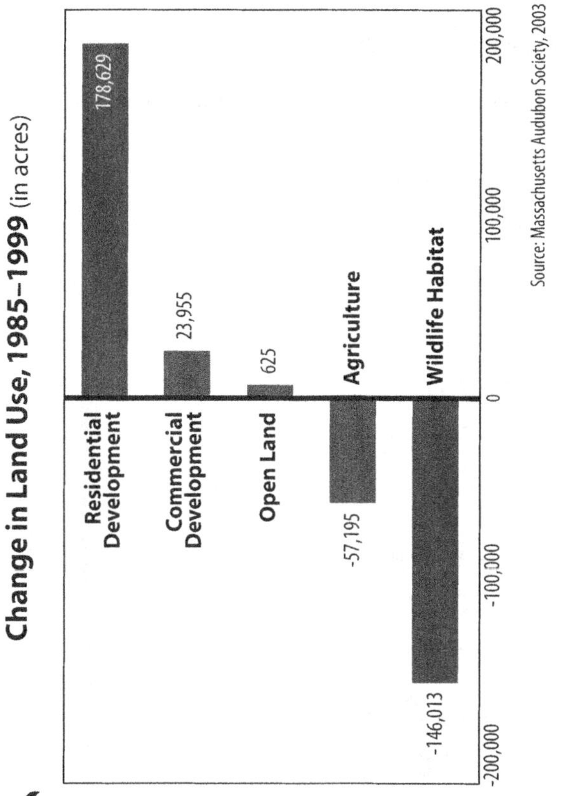
Change in Land Use, 1985–1999 (in acres)
Residential Development
Commercial Development
Open Land
Agriculture
Wildlife Habitat
178,629
23,955
625
-57,195
-146,013
200,000
100,000
0
-100,000
-200,000
Source: Massachusetts Audubon Society, 2003

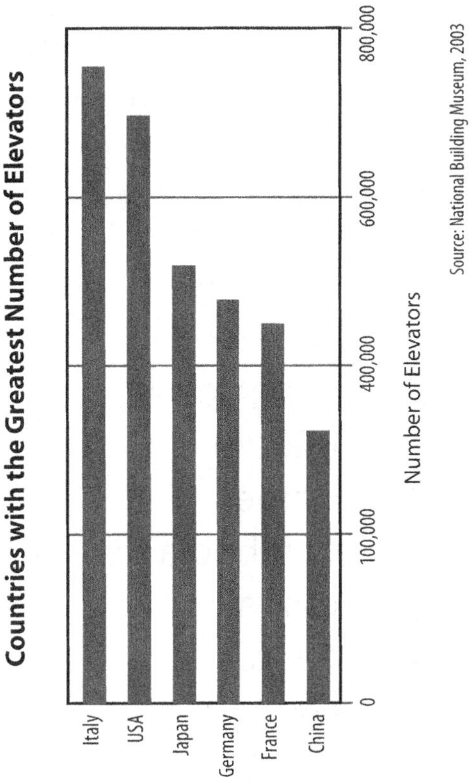
Countries with the Greatest Number of Elevators
Italy
USA
Japan
Germany
France
China
0
100,000
400,000
600,000
800,000
Number of Elevators
Source: National Building Museum, 2003

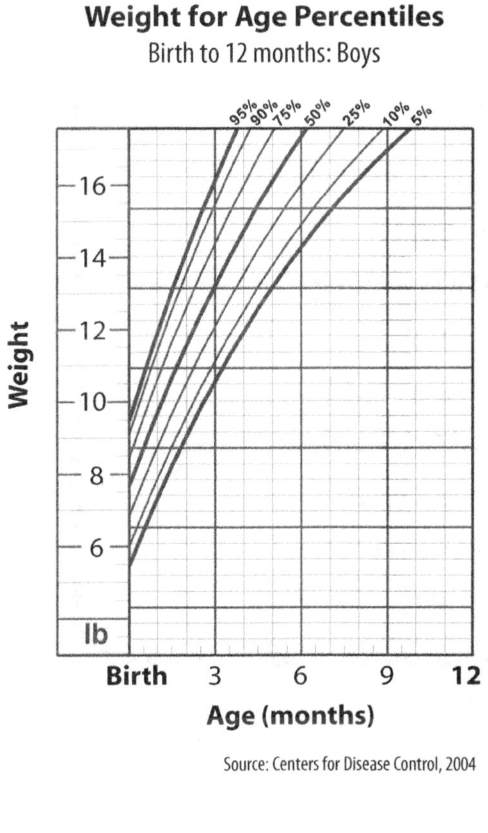
Weight for Age Percentiles
Birth to 12 months: Boys
95%
90%
75%
50%
25%
10%
5%
16
14
12
10
8
6
lb
Weight
Birth
3
6
9
12
Age (months)
Source: Centers for Disease Control, 2004

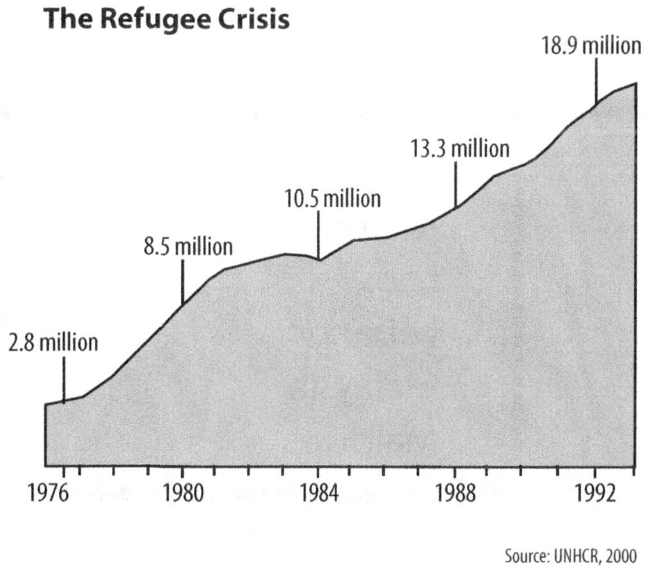
The Refugee Crisis
2.8 million
8.5 million
10.5 million
13.3 million
18.9 million
1976
1980
1984
1988
1992
Source: UNHCR, 2000

## Resource: Countries in Asia

| | |
|---|---|
| Afghanistan | Mongolia |
| Armenia | Myanmar |
| Azerbaijan | Nepal |
| Bahrain | North Korea |
| Bangladesh | Oman |
| Bhutan | Pakistan |
| Brunei | Palestine |
| Cambodia | Philippines |
| China | Qatar |
| Cyprus | Russia |
| Georgia | Saudi Arabia |
| India | Singapore |
| Indonesia | South Korea |
| Iran | Sri Lanka |
| Iraq | Syria |
| Israel | Taiwan |
| Japan | Tajikistan |
| Jordan | Thailand |
| Kazakhstan | Timor-Leste |
| Kuwait | Turkey |
| Kyrgyzstan | Turkmenistan |
| Laos | United Arab Emirates |
| Lebanon | Uzbekistan |
| Malaysia | Vietnam |
| Maldives | Yemen |

## Station 1: Ups and Downs 1

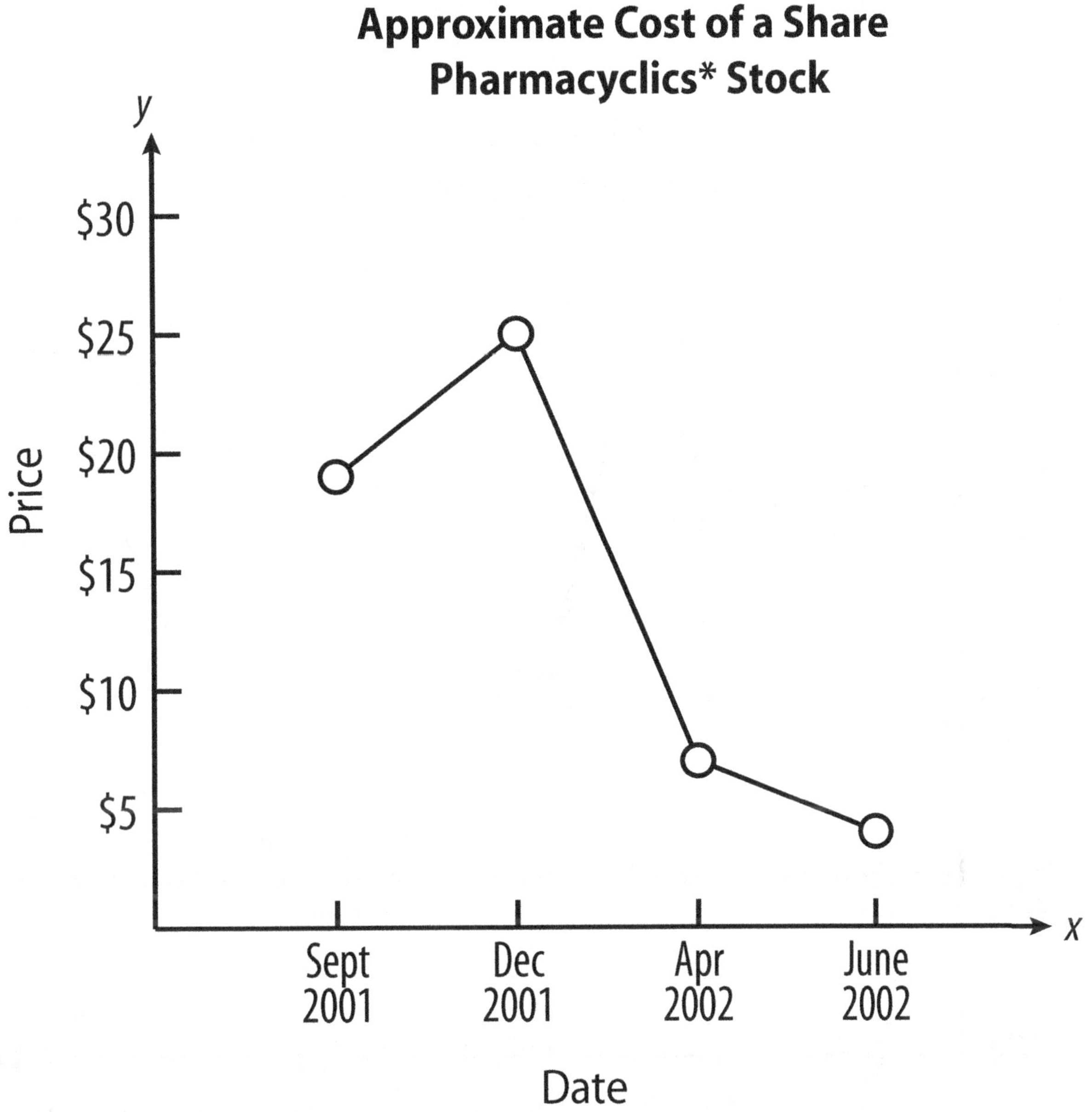


Source: http://www.BigCharts.com, 2003

* This is a cancer drug that did not meet the standards for FDA approval.

## Station 2: Ups and Downs 2

### Average Milk Prices per Gallon U.S. Cities, 2012 – 2022

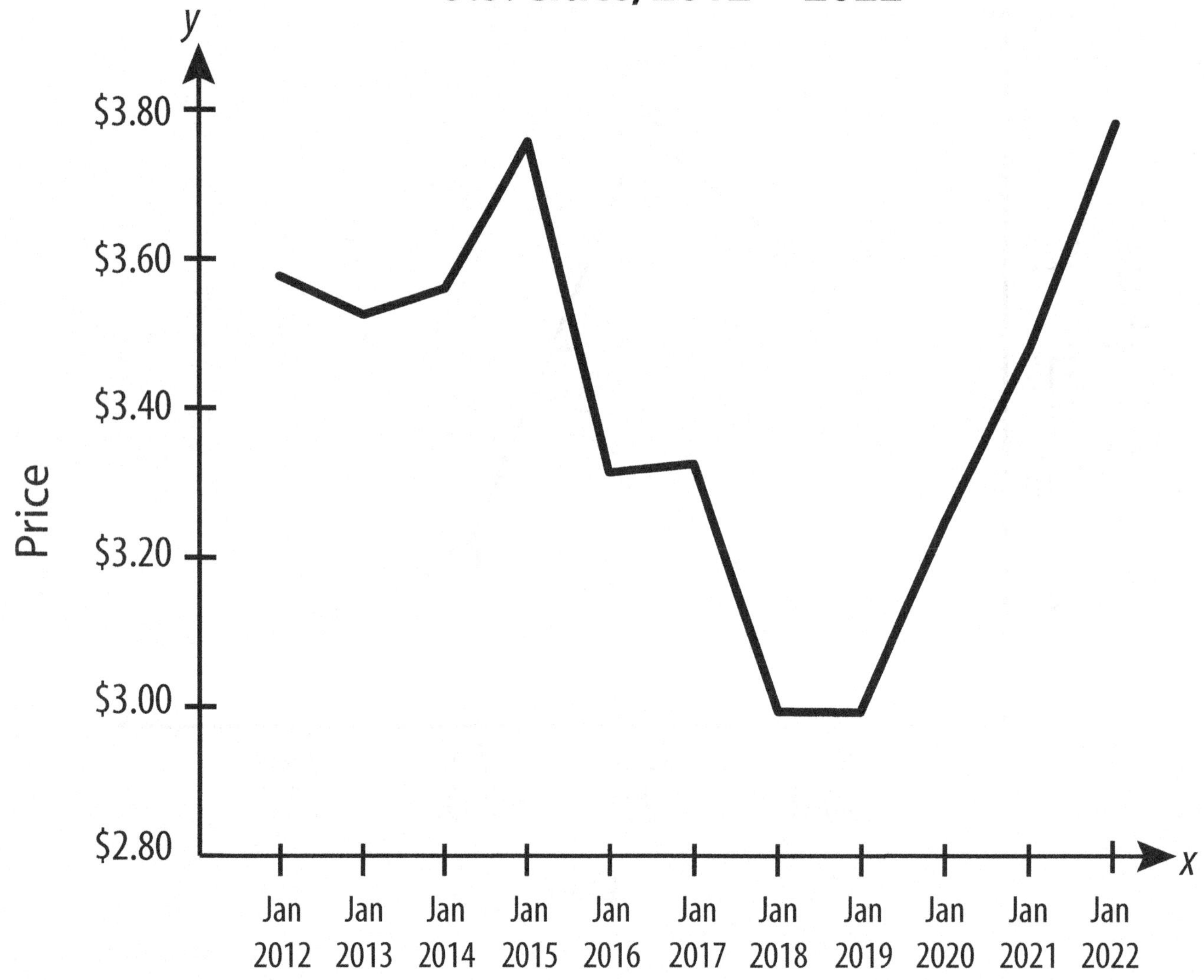

# Blackline Master 7

## Station 3: Jerri's Jobs

CORRECTED (if checked) ☐

PAYER'S name, street address, city or town, state or province, country, ZIP or foreign postal code, and telephone no.

**Jeff's Janitorial Service**
**192 Green Street**
**Augusta, ME 04330**

OMB No. 1545-0116

Form **1099-NEC**

(Rev. January 2022)

For calendar year 20 17

**Nonemployee Compensation**

PAYER'S TIN

RECIPIENT'S TIN

1 Nonemployee compensation
$ 10,675.45

**Copy B**

**For Recipient**

This is important tax information and is being furnished to the IRS. If you are required to file a return, a negligence penalty or other sanction may be imposed on you if this income is taxable and the IRS determines that it has not been reported.

RECIPIENT'S name
**Jerri XXX**

2 Payer made direct sales totaling $5,000 or more of consumer products to recipient for resale ☐

3

Street address (including apt. no.)

City or town, state or province, country, and ZIP or foreign postal code

4 Federal income tax withheld
$ 1,921.81

Account number (see instructions)

5 State tax withheld
$
$

6 State/Payer's state no.

7 State income
$
$

Form **1099-NEC** (Rev. 1-2022) (keep for your records) www.irs.gov/Form1099NEC Department of the Treasury - Internal Revenue Service

---

CORRECTED (if checked) ☐

PAYER'S name, street address, city or town, state or province, country, ZIP or foreign postal code, and telephone no.

**Jeff's Janitorial Service**
**192 Green Street**
**Augusta, ME 04330**

OMB No. 1545-0116

Form **1099-NEC**

(Rev. January 2022)

For calendar year 20 18

**Nonemployee Compensation**

PAYER'S TIN

RECIPIENT'S TIN

1 Nonemployee compensation
$ 4,950.00

**Copy B**

**For Recipient**

This is important tax information and is being furnished to the IRS. If you are required to file a return, a negligence penalty or other sanction may be imposed on you if this income is taxable and the IRS determines that it has not been reported.

RECIPIENT'S name
**Jerri XXX**

2 Payer made direct sales totaling $5,000 or more of consumer products to recipient for resale ☐

3

Street address (including apt. no.)

City or town, state or province, country, and ZIP or foreign postal code

4 Federal income tax withheld
$ 980.35

Account number (see instructions)

5 State tax withheld
$
$

6 State/Payer's state no.

7 State income
$
$

Form **1099-NEC** (Rev. 1-2022) (keep for your records) www.irs.gov/Form1099NEC Department of the Treasury - Internal Revenue Service

---

CORRECTED (if checked) ☐

PAYER'S name, street address, city or town, state or province, country, ZIP or foreign postal code, and telephone no.

**Jeff's Janitorial Service**
**192 Green Street**
**Augusta, ME 04330**

OMB No. 1545-0116

Form **1099-NEC**

(Rev. January 2022)

For calendar year 20 19

**Nonemployee Compensation**

PAYER'S TIN

RECIPIENT'S TIN

1 Nonemployee compensation
$ 0.00

**Copy B**

**For Recipient**

This is important tax information and is being furnished to the IRS. If you are required to file a return, a negligence penalty or other sanction may be imposed on you if this income is taxable and the IRS determines that it has not been reported.

RECIPIENT'S name
**Jerri XXX**

2 Payer made direct sales totaling $5,000 or more of consumer products to recipient for resale ☐

3

Street address (including apt. no.)

City or town, state or province, country, and ZIP or foreign postal code

4 Federal income tax withheld
$ 0.00

Account number (see instructions)

5 State tax withheld
$
$

6 State/Payer's state no.

7 State income
$
$

Form **1099-NEC** (Rev. 1-2022) (keep for your records) www.irs.gov/Form1099NEC Department of the Treasury - Internal Revenue Service

# Blackline Master 8

## Station 3: Jerri's Jobs (page 2)

☐ CORRECTED (if checked)

| | | | |
|---|---|---|---|
| PAYER'S name, street address, city or town, state or province, country, ZIP or foreign postal code, and telephone no.<br>**Jeff's Janitorial Service**<br>**192 Green Street**<br>**Augusta, ME 04330** | | OMB No. 1545-0116<br>Form **1099-NEC**<br>(Rev. January 2022)<br>For calendar year<br>20 **20** | **Nonemployee Compensation** |
| PAYER'S TIN | RECIPIENT'S TIN | **1** Nonemployee compensation<br>$ **5,600.00** | **Copy B**<br>**For Recipient** |
| RECIPIENT'S name<br>**Jerri XXX** | | **2** Payer made direct sales totaling $5,000 or more of consumer products to recipient for resale ☐ | This is important tax information and is being furnished to the IRS. If you are required to file a return, a negligence penalty or other sanction may be imposed on you if this income is taxable and the IRS determines that it has not been reported. |
| Street address (including apt. no.) | | **3** | |
| City or town, state or province, country, and ZIP or foreign postal code | | **4 Federal income tax withheld**<br>$ **1,008.21** | |
| Account number (see instructions) | | **5** State tax withheld<br>$<br>$ | **6** State/Payer's state no. · **7** State income<br>$<br>$ |

Form **1099-NEC** (Rev. 1-2022) (keep for your records) www.irs.gov/Form1099NEC Department of the Treasury - Internal Revenue Service

☐ CORRECTED (if checked)

| | | | |
|---|---|---|---|
| PAYER'S name, street address, city or town, state or province, country, ZIP or foreign postal code, and telephone no.<br>**Jeff's Janitorial Service**<br>**192 Green Street**<br>**Augusta, ME 04330** | | OMB No. 1545-0116<br>Form **1099-NEC**<br>(Rev. January 2022)<br>For calendar year<br>20 **21** | **Nonemployee Compensation** |
| PAYER'S TIN | RECIPIENT'S TIN | **1** Nonemployee compensation<br>$ **5,600.00** | **Copy B**<br>**For Recipient** |
| RECIPIENT'S name<br>**Jerri XXX** | | **2** Payer made direct sales totaling $5,000 or more of consumer products to recipient for resale ☐ | This is important tax information and is being furnished to the IRS. If you are required to file a return, a negligence penalty or other sanction may be imposed on you if this income is taxable and the IRS determines that it has not been reported. |
| Street address (including apt. no.) | | **3** | |
| City or town, state or province, country, and ZIP or foreign postal code | | **4 Federal income tax withheld**<br>$ **1,008.21** | |
| Account number (see instructions) | | **5** State tax withheld<br>$<br>$ | **6** State/Payer's state no. · **7** State income<br>$<br>$ |

Form **1099-NEC** (Rev. 1-2022) (keep for your records) www.irs.gov/Form1099NEC Department of the Treasury - Internal Revenue Service

Station 4: The Price of a Subway Ride

## Price of a Subway Ride

| Year | Cost |
|---|---|
| 1978 | $0.25 |
| 1980 | $0.50 |
| 1981 | $0.75 |
| 1982 | $0.60 |
| 1989 | $0.85 |
| 1999 | $1.00 |

# Blackline Master 10

# Blackline Master 11

## Blackline Master 12

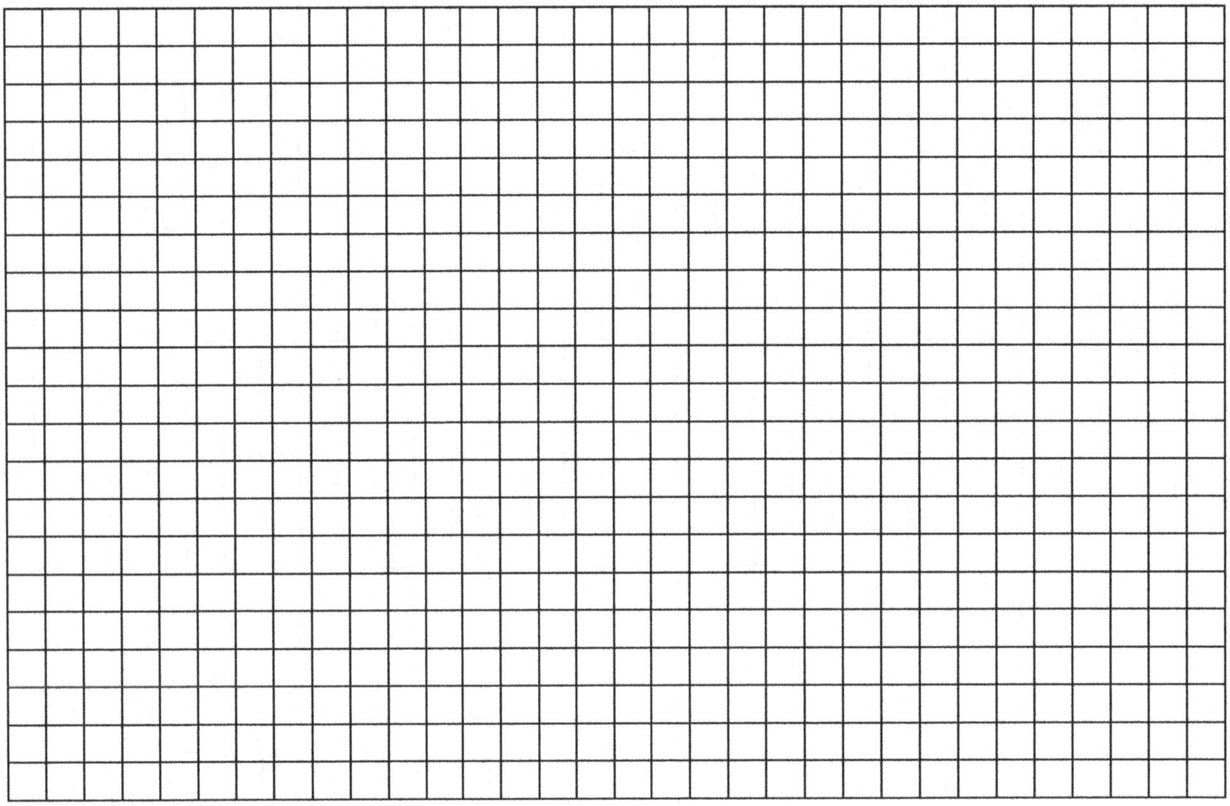

# Glossary

Reminder: Students generate their own definitions for terms as they arise in class, using language that makes sense to them. However, to help you guide the discussion, we include mathematical definitions for most of those terms.

Italicized terms are *not* included in the *Student Book*, but are included here as background information. If a term is introduced in a lesson, the lesson number is in parentheses following the term.

**Average** (7)—one of three things: It can describe the middle of the data set (median); the most typical value (mode); or the arithmetic average (mean).

**Bar graph** (3)—a graph that uses horizontal or vertical bars to show countable data; e.g., "Countries With the Greatest Number of Elevators, *BLM 3*."

**Benchmark fractions** (2)—fractions commonly used or used to find other fractions. They include $\frac{1}{2}$, $\frac{1}{4}$, $\frac{3}{4}$, and $\frac{1}{10}$.

**Benchmark percents** (2)—percent values commonly used or used to find other percents. They include 25%, 50%, 75%, 100%, and 10%.

**Category** (OU)—a subdivision of a larger group used to describe data. In the countries' data that were regrouped by continents, each continent is a category.

**Change over time** (5)—the change that the numerical value of a particular thing goes through over a given time period. For example, examining the attendance at parks over a 20-year-period will show the change in attendance over time.

**Circle graph** (3)—a graph showing parts of the data related to the whole and to each other; also known as a pie graph or pie chart.

**Criteria** (IA)—a standard against which individuals or things are evaluated, for example, criteria for selecting high-level responses.

**Climate** (8)—surrounding conditions affecting ambiance, atmosphere, environment; in the case of *Mystery Cities*, the conditions of rain (precipitation) and temperature.

**Climate graph** (8)—a graph that displays data for one or more aspects of weather; e.g., temperature, precipitation, relative humidity, or average wind speed.

**Data, data set** (1)—a set of information that can be numerical (e.g., salaries, test scores, heights, age) as well as categorical (e.g., countries, foods we eat, types of product defects). One piece of information is called datum.

**Frequency graph** (1)—a graph showing a tally for each category or group; it has three components: a line, tallies, and labels.

**Graph** (OU)—a representation of information that may include either individual points, bars, lines, or circles.

**Horizontal or *x*-axis** (5)—the horizontal (side to side) number line on a coordinate plane.

**Interval (6)**—the set of numbers between two numbers, used in graphs to show data; for instance, between 1980 and 1990 the interval is 10 (years).

**Key** (3)—a list of words or phrases giving an explanation of symbols or abbreviations. *Synonym*: legend.

**Legend** (3)—an explanatory list of the symbols on a map or chart. *Synoym*: key.

**Line graph** (5)—a graph that uses a line to show how data change over time.

**Mean** (7)—the arithmetic average of a set of data. Used to find grades in school; e.g., if test grades are 73, 58, 92, and 79, the mean is 302 (the sum) divided by 4 (the number of tests), or 75.5.

***Measure of Central Tendency***—an average. There are three types of measures of central tendency: mean, median, and mode.

**Median** (9)—the middle number or the average of two middle numbers in an ordered set of data. The median is the point below which half of the data fall. For example, the median of 2, 3, 5, 7, and 8 is 5, and the median of 2, 3, 5, 7, 8, and 9 is 6. When a data set includes extreme outliers, the median is often the best indicator of center or middle.

**Mode** (1)—the data point (individual number or category) that appears most often in a set of data. For example, in the Springfield data set, the mode is the United States. The United States is the country (category) where most of the clothes were made.

***Outliers*** (7)—values at the extreme ends of a data set. In this example of scores—89, 10, 96, 92, and 100—10 could be considered an outlier.

***Population*** (1)—the entire collection of items that is the focus of concern.

**Precipitation** (8)—amount of rainfall.

***Random sample***—a set of items drawn from a population in such a way that each time an item is selected, every item in the population has an equal opportunity to appear in the sample.

**Representative sample** (2)—a sample that includes careful selection across the population being surveyed.

**Sample** (1)—the data that are collected; e.g., the countries where students' clothes are made.

**Scale** (6)—on a graph, the numbers placed at fixed distances to help label the graph; e.g., a scale of 2 is used in a graph showing the years 1995, 1997, 1999, etc.

**Statistic** (1)—a numerical value used to describe data; e.g., the percent of marriages in the United States that ended in divorce in 2000.

**Stocks** (10)—capital funds of a company held jointly by its owners.

**Shares** (10)—part ownership of the company.

**Stock market**(10)—the business of buying and selling shares of stocks in companies.

**Table** (4)—a display of data using a chart with labels and counts.

**Trend** (6)—a continuing change in the same general direction; e.g., amusement park attendance shows a trend.

**Vertical axis or *y*-axis** (3)—the vertical (up and down) axis on a coordinate plane.

# Answer Key

## Opening the Unit

**Activity 1: Making a Mind Map**

Answers will vary.

**Activity 2: Initial Assessment**

Refer to the *Initial Assessment Checklist*, p. 146, and *Looking Closely*, p. 5, for suggestions when observing students.

**Task 1: Decisions**

1. Answers will vary. Sample answer: Where to take my nephews when they come here for a visit.
2. Answers will vary. Sample answer based on above sample: Important criteria to consider are my nephews' interests, the cost involved, and the travel time needed.
3. Answers will vary. Sample answer based on above sample: I will need to contact my nephews and ask them what they like to do. I will check online to find out the costs of places of interest in the area. I will check a map to see about how far it is to drive to the various places of interest.
4. Answers will vary. Sample answer based on above sample: I'll make a table with three columns once I figure out what places might interest my nephews:

| Place of interest | Estimated cost involved | Driving time |
|---|---|---|
| | | |

**Task 2: Printing Plant**

1. The printing plant mistakes are mostly caused by the paper cutter.
2. a. Printing Plant Errors: Machine vs. Human

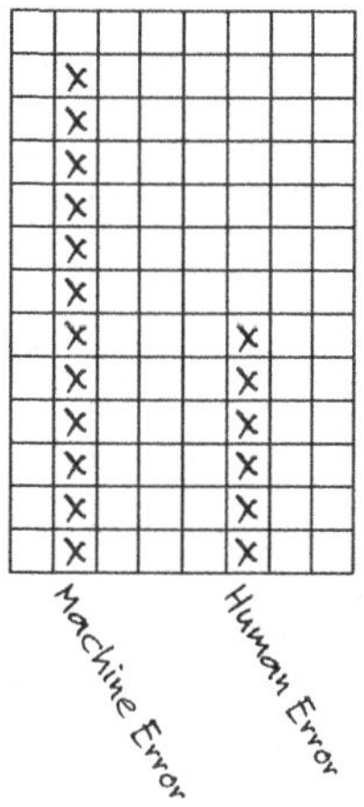


b. Printing Plant Errors: Machine vs. Human

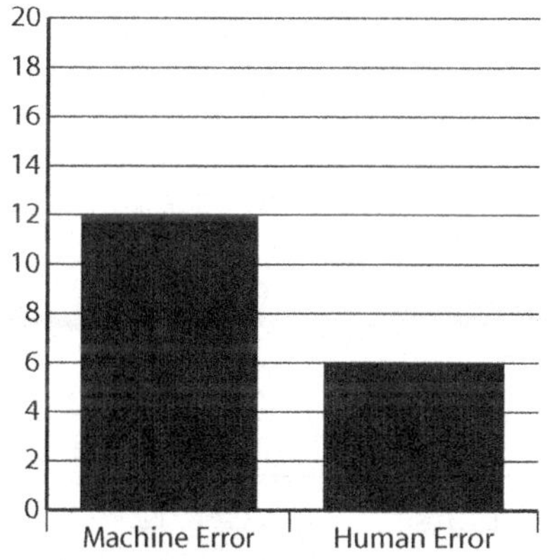


c. Printing Plant Errors: Machine vs. Human

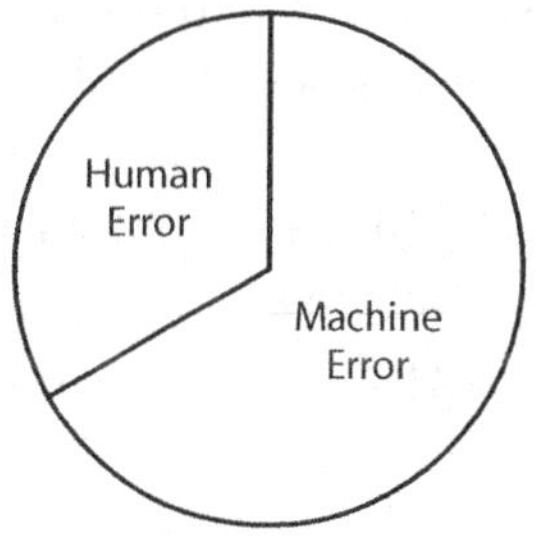

3. Answers will vary. Sample answers: There are twice as many machine errors as human errors. The human errors represent one-third of the total number of errors.
4. Answers will vary. Sample answer: The bar graph shows very readily how the two categories compare, and it is easy to figure out exactly how many of each error were reported.
5. Answers will vary. Sample answer: Rosa could try checking out the paper cutter to see why it is making so many errors. It might need cleaning or readjusting or even replacing.

**Task 3: Stocks, Tables, and Line Graphs**

1. Answers will vary. Sample answer: The stock began at $45.00 and rose sharply the first day, then continued to progress but at a much slower pace.
2.

| Day | Price of Share |
|---|---|
| W | $45.00 |
| T | $50.50 |
| F | $51.00 |
| T | $51.00 |
| W | $52.00 |

3.

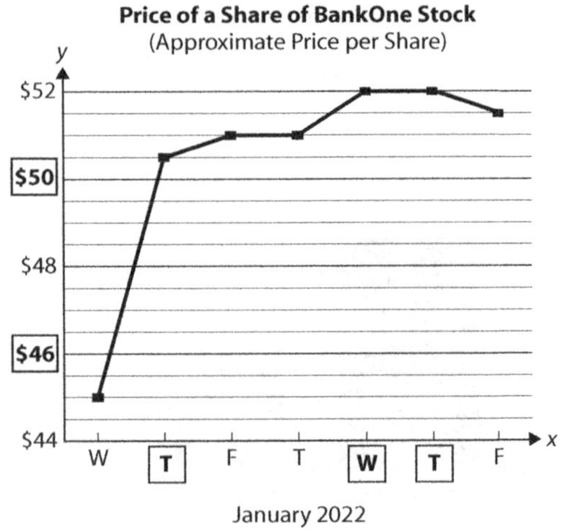


4. See Question 3 above.
5. If you wanted to see the ups and downs more clearly, you would need a scale of at least **$5 per interval**.
6. a. The mean is **$49.90** ($50.43 if two added data points are included).
   b. The median is **$51.00** (with or without the two added data points).
   c. Answers will vary. Sample answer: The median seems to more closely reflect the data since there are two data points on $51.00, and all the others but one ($45.00) hover just below and just above it.

**Activity 3: Sorting Graphs**

1. Answers will vary but should reflect a reasonable organization strategy. Sample answer: The graphs are sorted into three piles:

   Circle graph pile, which includes "Smoking-related Deaths," "Donors' Blood Types," and "The Changing Needs of the American Workforce."

   Bar graph pile, which includes "Fatal Occupational Injuries 1990–2000: Logging vs. Grocery Store Work," "Change in Land Use, 1985–1999," "Countries with the Greatest Number of Elevators," and "U.S. Population by Age."

   Line graph pile, which includes "Unemployment Rate: 2008–2018," Weight for Age Percentiles," and "The Refugee Crisis," "Northern Hemisphere Temperature 1400–2000," and "Immigration from China 1850–1900."
2. Answers will vary. Sample answer: All the graphs in the circle graph pile are round. Though only 2 of the 3 sets of circle graphs actually use percents, they all use "slices" to represent the size of the parts compared to the total amount.

   All the graphs in the bar graph pile use bars to show either total amounts or percents for each category represented.

   The graphs in the line graph pile use lines to represent change over time.
3. Answers will vary. Sample answer: Although all the graphs in the circle graph pile are round, not all use percents. "Smoking-Related Deaths" uses actual numbers rather than percents. For the other sets of circle graphs ("Donors' Blood Types" and "The Changing Needs of the American Workforce"), you cannot figure out what the actual amounts are since you don't know the total of each category. All the graphs in the bar graph pile use bars to show either total amounts or percents for each category represented. The graphs in the line graph pile use lines to represent change over time.

   Some of the bars in the bar graph pile are vertical (up and down) and some are horizontal. One ("Change in Land Use, 1985–1999") shows both positive and negative amounts. The bars each represent something different: percents, number of fatalities, number of elevators, and acres of land.

The line graphs include straight lines from one point to another, but there is one that has "squiggly" lines without clearly defined points ("Northern Hemisphere Temperature 1400–2000"). One of the graphs in the line graph pile actually has several sets of lines representing percents ("Weight for Age Percentiles"). In one graph ("The Refugee Crisis") the area below the line graph is shaded in.

### Practice: The Graph I Chose

1. Answers will vary. See *Assessing Written Work: The Graph I Chose*, p. 12, for evaluating student answers.
2. Answers will vary. See *Assessing Written Work: The Graph I Chose*, p. 12, for evaluating student answers.

### Practice: Data Collection

Answers will vary.

## Lesson 1: Countries in Our Closets

### Activity 1: Organizing the Data

See *Looking Closely*, p. 21, for suggestions on what to look for as students organize their data.

### Activity 2: Statements About Data

1. There are **three** times as many clothes from South Africa as from Barbados.
2. The number of clothes from Mexico is **one-third** of the number of clothes from China.
3. **Twice** as many clothes come from the United States as from China.
4. The clothes from Korea are **one-fourth** as many as those from Japan.
5. **Six** countries have only one article of clothing.
6. **Half** of the clothes come from **China** and **the United States**.
7. Japan has **four** times as many clothes as Germany.

### Activity 3: Changing the Categories

1. Students should notice that there are fewer categories since countries have been lumped into continents.
2. Students should notice that the story of the new graph (by continent) is less detailed; that is, it is no longer possible to tell which country within a continent produced more clothes.
3. Answers will vary.
4. Answers will vary but should match the information in the graph.

### Practice: Clothes by Continent in Springfield, MA

1.

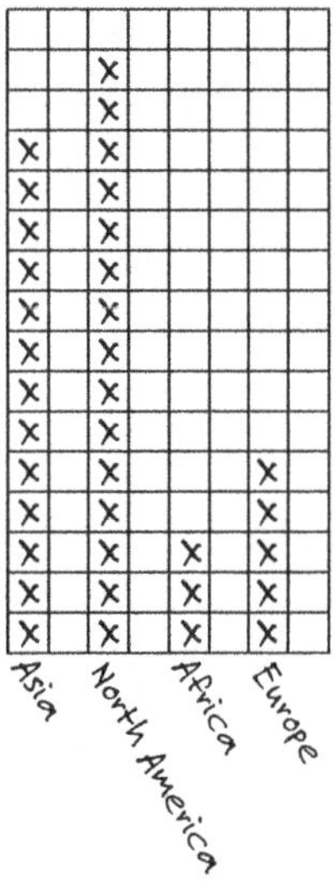


2. Answers will vary but should match the information in the graphs.
3. Answers will vary. Sample statements may be similar to the following:

   There are more clothes from North America than from any other continent.

   There are almost as many clothes from Asia as there are from North America.

   There are no clothes made in South America.

   Less than one-fourth of the clothes come from Africa and Europe.

   Over three-fourths of the clothes come from the two continents of Asia and North America.

### Practice: Reporting Data 1

1. Answers will vary. Sample statements may be similar to the following:

   There are more White AIDS patients than any other race/ethnicity.

   White and African American patients represent three-fourths of all the AIDS patients.

2.

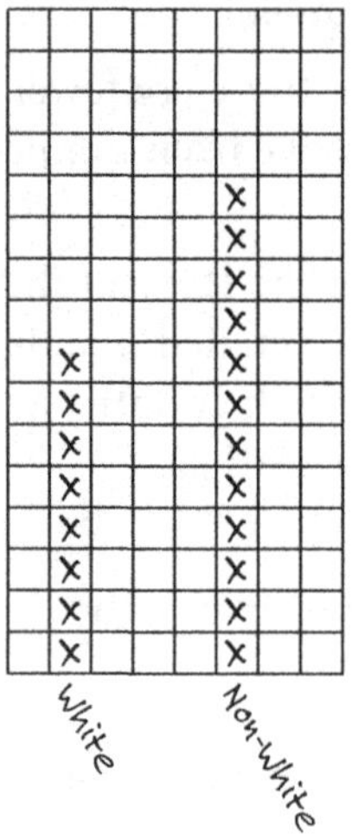


3. Sample response: With only two categories, there are fewer Whites than non-Whites.
4. Answers will vary.
5. Answers will vary.

**Practice: Reporting Data 2**

Sample regrouping:

| Travel Time to Work | Percent of Commuters | Number Based on 25 People |
|---|---|---|
| Less than 10 minutes | 16% | 4 |
| Between 10 minutes and 29 minutes | 52% | 13 |
| At least 30 minutes | 32% | 8 |

1. What is the travel time category with the biggest percent of commuters?

   Five categories **10–19 minutes**

   Three categories response based on above sample regrouping: **10–29 minutes**
2. Which category has the smallest percent of people?

   Five categories **45 minutes or more**

   Three categories response based on above sample regrouping: **Less than 10 minutes**
3. Answers will vary, depending on regrouping of data.
4. Answers will vary.

**Practice: Categorically Speaking**

Answers will vary.

**Extension: Taking Inventory**

See *Looking Closely*, p. 21, for evidence of student performance.

**Test Practice**

1. (1)
2. (4)
3. (2)
4. (4)
5. (2)
6. 58.6%

## Lesson 2: Most of Us Eat...

**Activity 1: Frequently Eaten Foods**

See *Lesson 2 in Action*, p. 38, for suggestions on what to look for as students organize their data.

**Activity 2: Consistent Categories**

1. Answers will vary. Sample answer: Elbow, because it is not a part of the hand.
2. Answers will vary. Sample answer: Early morning, because it is a more specific description of part of the day, where the other three are more general times of the day.
3. Answers will vary. Sample answer: Sales tax, because it is a more general term that includes all the other items in the category.
4. Answers will vary. Sample answer: Footwear, because it is a more general term that includes all the other items in the category.
5. Answers will vary. Sample answer: Dietary Preferences.
6. Answers will vary. Sample answer: Catastrophic Events.
7. Answers will vary. Sample answer: Seeking Employment.
8. Answers will vary. Sample answer: The Growth of Plants.

**Activity 3: Describing the Data**

Answers will vary but should match the information in the graph.

**Activity 4: Who to Ask?**

1. (2)
2. (2)

3. (3)
4. (1)

**Practice: Take a Sample**

1. Answers will vary. Sample answer: Freda could ask customers for feedback when they come to her ice cream shop.
2. Answers will vary. Sample answer: They could survey customers who have used their service department.
3. Answers will vary. Sample answer: Teachers could randomly call some of the students who have dropped out of the program.

**Practice: Thirsty**

1. Answers will vary but should include all the information provided. Sample answer:

| Juices | Carbonated Beverages | Caffeinated Beverages | Other |
|---|---|---|---|
| Orange Juice | Cola | Tea | Milk |
| Apple Juice | Soda | Tea | Water |
| Apple Juice | Ginger ale | Iced tea | Water |
| Cranberry Juice | | Coffee | Water |
| Cranberry Juice | | | Beer |

2. Answers will vary. See table above for sample titles.
3. Answers will vary. Sample answer: Juices—5; Carbonated beverages—3; Caffeinated beverages—4; Other—5.
4. Answers will vary. Sample answers: There are fewer carbonated beverages than any other type of drink. Juices represent less than one-third of the total number of drinks.
5. Answers will vary. Sample answers: Most of the adults said that they prefer juice over any of the other categories of drinks. Only a few said that they preferred a carbonated drink. None said that they preferred carrot juice.
6. Answers will vary. Sample answer: I would bring a variety of bottled juices and bottled water and then have a pot of hot water for tea. That would take care of at least 11 of the 17 students' preferences.

**Practice: Ouch!**

1.

| Contagious: 7 | Noncontagious: 21 | |
|---|---|---|
| Stomach flu | Numbness in hand | Backache |
| Cold | Migraine | Fever |
| Cold | Migraine | Bronchitis |
| Flu | Cramps | Hangover |
| Flu | Toothache | Insomnia |
| Flu | Allergy attack | Asthma |
| Chicken pox | Arthritis | Asthma attack |
| | Arthritis | Shortness of breath |
| | Sprained wrist | Heart attack |
| | Sprained ankle | Headache |
| | Backache | |

Answers will vary. Sample answer:

| Possibly Work-Related: 5 | Not Work-Related: 14 | Unclear: 9 |
|---|---|---|
| Numbness in hand | Stomach flu | Insomnia |
| Sprained wrist | Cold | Asthma |
| Sprained ankle | Cold | Asthma attack |
| Backache | Flu | Shortness of breath |
| Backache | Flu | Heart attack |
| | Flu | Headache |
| | Cramps | Migraine |
| | Toothache | Migraine |
| | Arthritis | Allergy attack |
| | Arthritis | |
| | Fever | |
| | Bronchitis | |
| | Hangover | |
| | Chicken pox | |

2. a. Answers will vary. See tables above for suggested labels.
   b. Contagious—7, Noncontagious—21; Possibly work-related—5; Not work-related—14; Unclear—9.
   c. Answers will vary. Sample answer: There are three times more noncontagious than contagious health issues. The contagious issues represent one-fourth of the total health issues represented.

There are almost three times as many non-work-related health issues as possibly work-related issues. The non-work-related issues represent one-half of the total health issues represented.

3. a. Answers will vary. Sample answer: Most of the health problems were noncontagious issues. Most of the health problems were not work-related.
   b. Answers will vary. Sample answer: None of the health problems were untreatable.

**Extension: Friend and Family Drink Too**

1. Answers will vary.
2. Answers will vary.
3. Answers will vary.
4. Answers will vary.
5. Answers will vary.
6. Answers will vary.

**Extension: Data Collection**

The data collected will vary.

**Test Practice**

1. (1)
2. (3)
3. (2)
4. (3)
5. (2)
6. $2,626

## Lesson 3: Displaying Data in a New Way

**Activities 1 and 2: Constructing a Bar Graph and Constructing a Circle Graph**

See *Lesson 3 in Action*, p. 47, for sample student work.

**Practice: Label That *Y*-Axis**

Entries for the *x*-axes may vary.

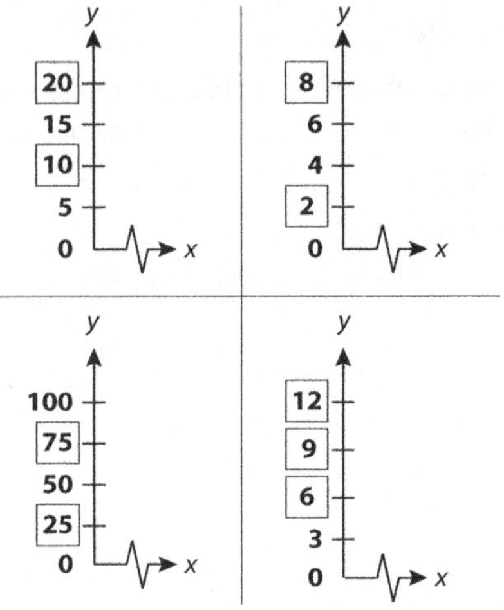


**Practice: Bar Graph Details**

1. The labels are 5, 10, 15, 20, and 25.
2. From the information on the graph, we can say that combined, the two biggest metro areas in Texas have a **smaller** population than New York City. Houston and **Washington, DC** combined have about the same population as Los Angeles.

3a. The first graph shows that **Italy** has the greatest number of elevators; almost **800,000** elevators.

3b. **USA** has the second greatest number of elevators. However, in the second graph, it appears that this country does not have as many elevators per 10,000 people as do **Italy**, **Japan**, **Germany**, or **France**.

4. Sample response: There aren't many tall buildings in China.

**Practice: Bar Graph to Circle Graph**

**Worldwide Internet Browser Market Share, May 2022**

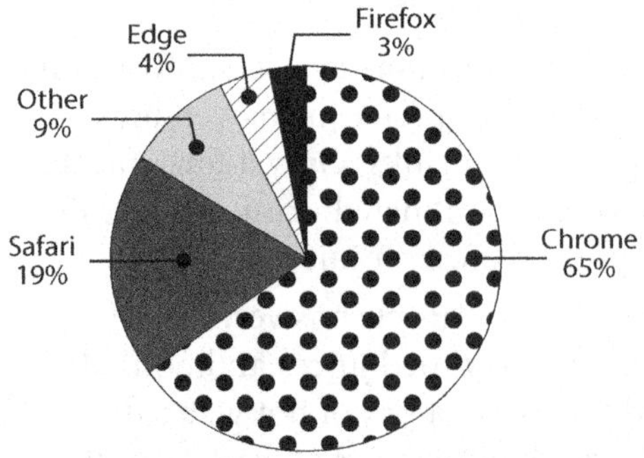

### Extension: Foods You Eat in a Week

See *Looking Closely*, p. 43, for evidence of student performance.

### Extension: Net Gains and Losses

1. Net gains and losses of land use
2. 146,013 acres lost
3. Wildlife habitat. It is the longest bar moving towards the left.
4. Residential development. It is the longest bar moving towards the right.
5. Answers will vary. You may want to check the student's original data with those presented in the graph.

### Test Practice

1. (2)
2. (2)
3. (2)
4. (4)
5. (3)
6. 93%

## Lesson 4: A Closer Look at Circle Graphs

### Activity 1: How Many of Each?

See *Looking Closely*, p. 55, for suggestions on what to look for as students organize their data and sketch their work.

### Activity 2: Cheat Sheet

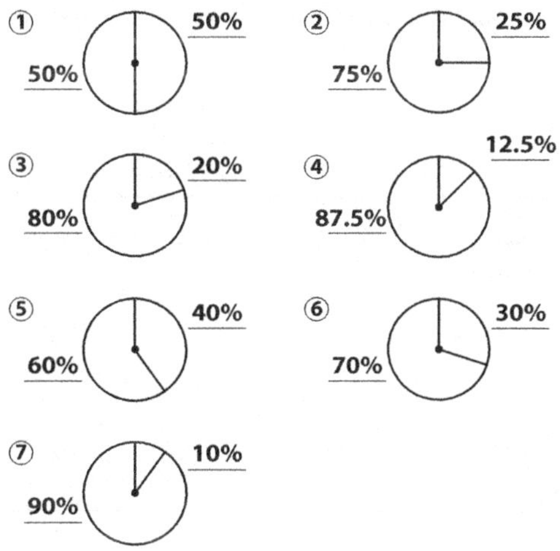


### Activity 3: Three Possible Answers

1. Answers will vary, but graphs should be representational of the information collected.
2. Answers will vary, but graphs should be representational of the information collected.
3. a.

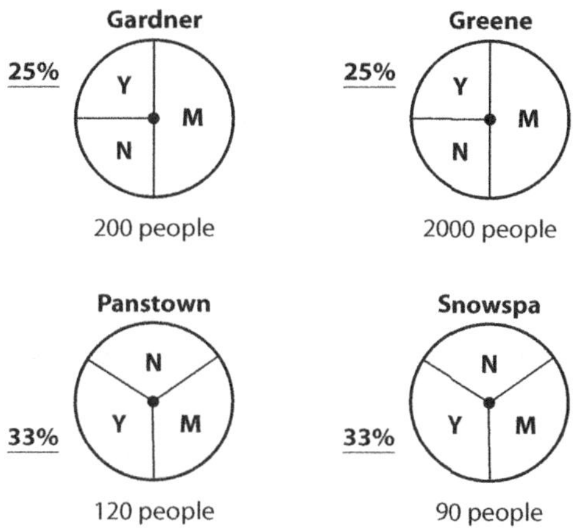


   b. Greene had the most yes votes. 25% of 2000 = 500.

   c. Snowspa had the least yes votes. 33% of 90 = 30.

   d. Answers will vary but should be based on determining percent of the total.

   e. Answers will vary.

### Practice: Estimating Percents and Parts of the Whole

Answers may vary slightly from these estimates.

1. A is about 55%.
   B is about 15%.
2. A is about 45%.
   C is about 30%.
3. A is about 50%.
   B is about 30%.
4. A is about 50%.
   B is about 40%.
5. 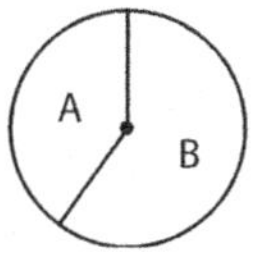

6. 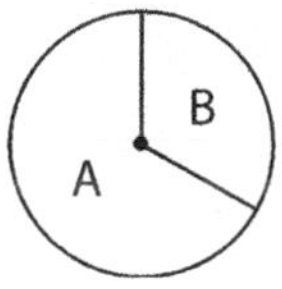

7.

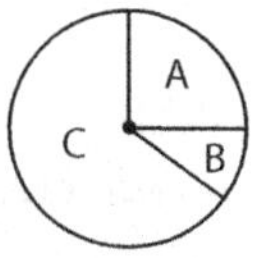


8.

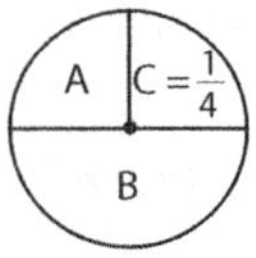


**Practice: Serve Up the Pie (Graphs)**

1.

| Flavor | How Many | Fraction of Total | Benchmark Percent |
|---|---|---|---|
| Chocolate | 52 | 52/102 | 50% |
| Vanilla | 26 | 26/102 | 25% |
| Strawberry | 24 | 24/102 | 25% |
| Total | 102 | 102/102 | 100% |

**Favorite Ice Cream Flavors**

2.

| Defect Type | How Many | Fraction of Total | Benchmark Percent |
|---|---|---|---|
| Broken Coil | 26 | 26/50 | 50% |
| Loose Pin | 13 | 13/50 | 25% |
| Missing Part | 11 | 11/50 | 25% |
| Total | 50 | 50/50 | 100% |

**Quality Control**

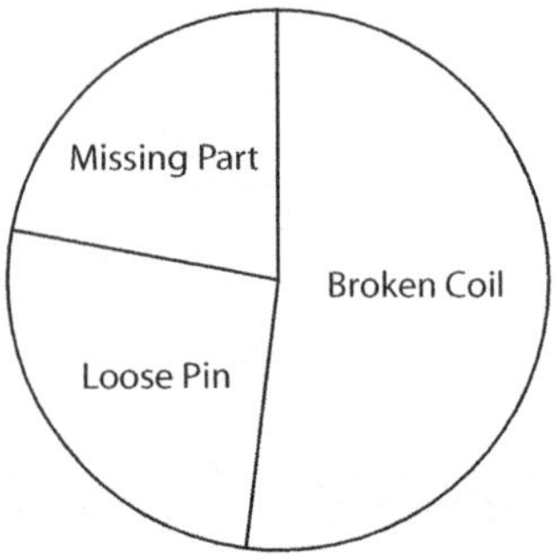


3.

| Affiliation | How Many | Fraction of Total | Benchmark Percent |
|---|---|---|---|
| Democrat | 99 | 99/200 | 50% |
| Republican | 47 | 47/200 | 25% |
| Independent | 54 | 54/200 | 25% |
| Total | 200 | 200/200 | 100% |

**Registered Voters**

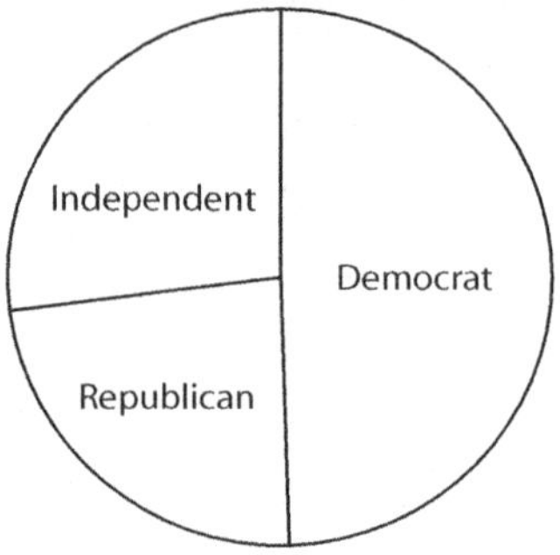


**Practice: Making Predictions**

1. Which color would you recommend she order the most of? **Black**
2. Which color might she consider not buying? **Red**

If you were to pick one candy out of the bag without looking, which color would you most likely pick? _ **Yellow**

If you picked a candy out of the bag without looking, which color would you be least likely to pick? **Brown**

**Extension: The Computer Makes the Graph**

Answers will vary, depending on how the data was collapsed into three categories.

Sample answer:

| Through Age 24 | Ages 25–54 | Ages 55 and Older |
|---|---|---|
| 35.3% | 43.6% | 21% |

**U.S. Population by Age**

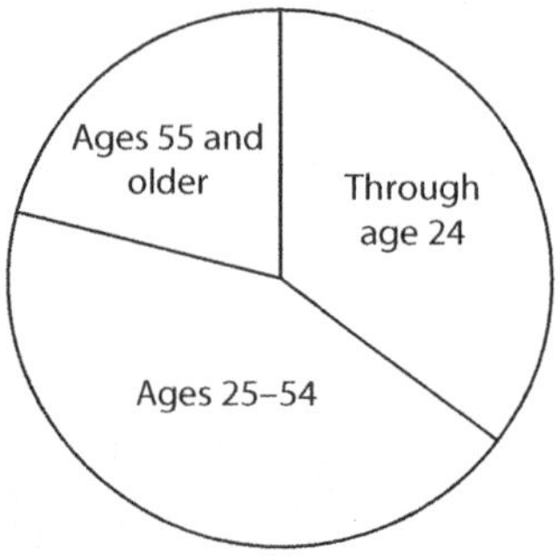

### Extension: Search for Circle Graphs

Answers will vary.

### Test Practice

1. (2)
2. (4)
3. (5)
4. (2)
5. (4)
6. 9,600

## Midpoint Assessment

### Decisions with Data

Answers will vary. Sample answer: The reputation of the landlord and the location are important to me. I would go to the apartment building and interview some of the people I saw hanging around. That way I could find out about the landlord and check out the location at the same time.

Answers will vary. Sample answer: Kindness and honesty are two important criteria in choosing a new partner. I would hang around the person and listen and observe closely to find out just how honest he is. I would watch how that person treats other people and small animals and how he talks about people less advantaged than he is.

### Activity: The Data Say

See *Assessment Activity: The Data Say*, p. 63, and *Looking Closely*, p. 65, for suggestions on what to observe as students work through their activities.

### Activity: Option 1, A Look at Your Program

Answers will vary, but the graph created should reflect the data students collected.

### Activity: Option 2, Your Survey

Answers will vary, but the graph and summary statements should reflect the data students collected.

### Activity: Option 3, Crime Fighter

Answers will vary. Sample answer:

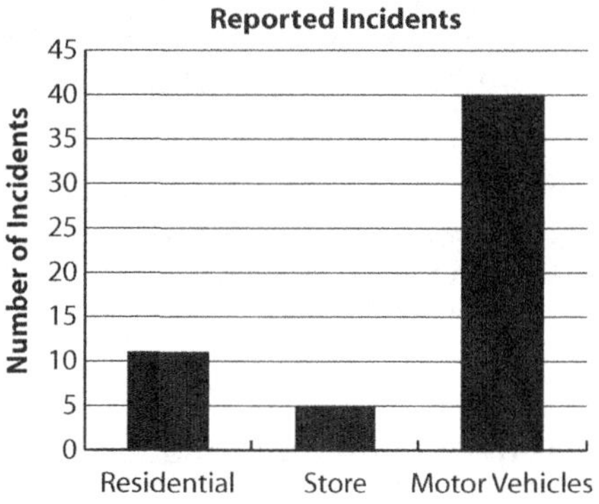


When I talked with the Crime Watch officer, I would want to ask him about watching vehicles more closely. Of the 56 incidents reported, 40 of them involved motor vehicles: 12 were stolen vehicles, 22 were vehicular break-and-enters, and 6 were larcenies from vehicles. There were 5 store robberies, which represent about $\frac{1}{10}$ of the total number of incidents reported. In 11 out of 56 incidents, residential break-ins occurred, about $\frac{1}{5}$ of the total. The incidents involving motor vehicles were well over 50% of the total number, so this should be the area of greatest focus.

### Activity: Option 4, Eating Out

Answers will vary. Sample answer:

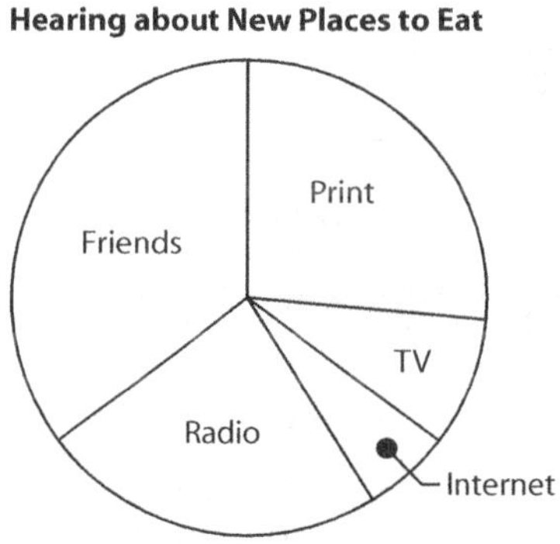


Based on the data, I would tell the restaurant owner to be sure that he or she serves great food at reasonable prices and doesn't make customers wait too long. Those are points that people mention to their friends. Because one-third of the people heard about new places to eat from their friends, the owner should be sure that people have good experiences to share. I would also tell the owner to put some special deal coupons in the newspapers or magazines because 9 out of 34 people (more than 25%) said they hear about new places to eat through those print materials.

### Practice: Decisions, Decisions

Answers will vary. Sample answers:

**Choosing a School**

| Criteria | Where to find information |
|---|---|
| Financial aid available | Go to the business office of the school. |
| Location | Check online to find out where the schools are located; if possible, drive there. |

**Choosing a Health Plan**

| Criteria | Where to find information |
|---|---|
| Cost | Look at their available print materials. |
| Types of services offered | Look at their available print materials; talk to a representative. |

**Choosing a Mortgage Company**

| Criteria | Where to find information |
|---|---|
| Willing to do business with me | Call different mortgage companies to see whom they serve. |
| Reputation | Check with friends; check online to see how often the companies have sold their mortgages. |

**Choosing a Bank**

| Criteria | Where to find information |
|---|---|
| Willing to do business with me | Go to the different banks and see which seems most accessible to driving. |
| Services offered | Look at their brochures; check online to see what services they provide. |

**Choosing a Lawyer**

| Criteria | Where to find information |
|---|---|
| Experience with type of case | Look in *Yellow Pages*, check online, and then ask friends. |
| Rates | Call lawyers' offices to get estimates; talk to friends |

**Choosing a Microwave**

| Criteria | Where to find information |
|---|---|
| Features/ Flexibility | Go to different stores and look at the different types; check online to see what's available. |
| Warranty | Check online to see what is available with each model; ask salespeople at various stores selling microwaves. |

**Choosing the Timing for Having a Medical Procedure**

| Criteria | Where to find information |
|---|---|
| Seriousness of procedure | Ask doctor; read up on the procedure online; ask friends. |
| Other events occurring in my life | Look at calendar to see whether there are any special events planned (like weddings), then plan around those events. |

**Choosing a Babysitter**

| Criteria | Where to find information |
|---|---|
| Reputation | Ask friends and neighbors for recommendations. |
| Age/ Experience | Ask potential babysitters; ask friends and neighbors. |

## Lesson 5: Sketch This

### Activity 1: Sketch This

See *Looking Closely*, p. 71, for suggestions on what to look for as students sketch their graphs.

### Practice: Candy Bars

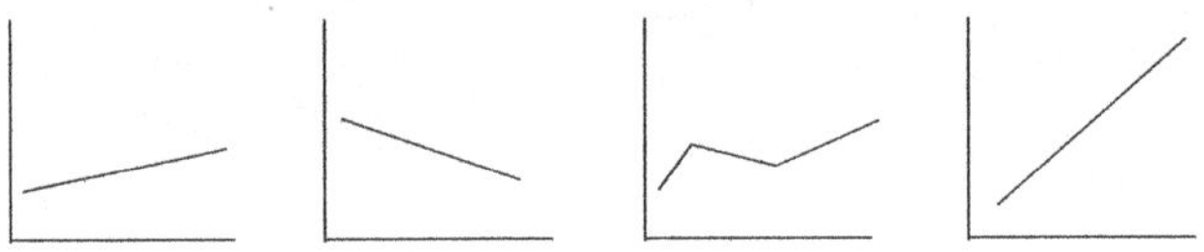

D B C A

Answers will vary, but statements should include clues such as "risen dramatically," "decreased," "fluctuated," and "slowly but steadily."

Answers will vary, but graph should reflect the statement describing the cost of a candy bar.

### Practice: How Does It Go?

Answers will vary, but sketches should be somewhat similar to the following:

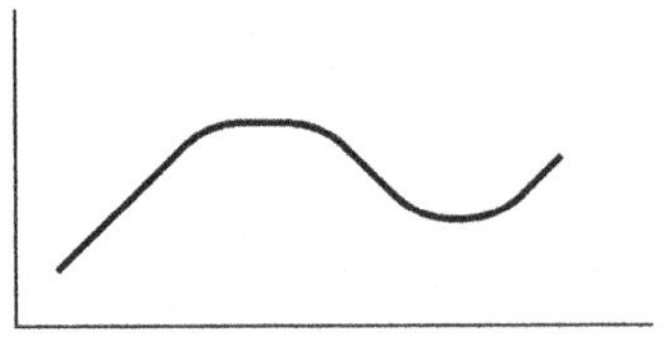

### Practice: Looking at the Entire Graph

1. B
2. A

### Practice: Labeling Graphs

1.

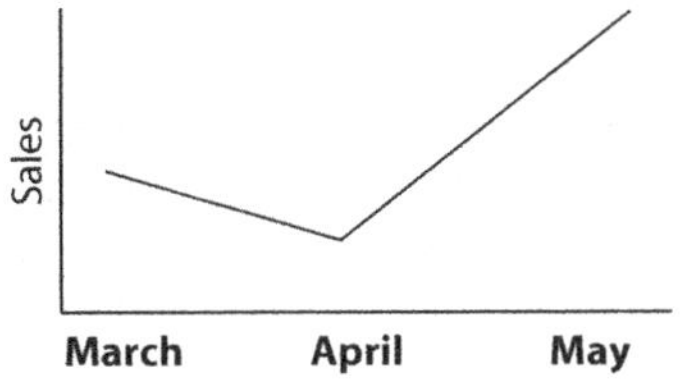


2.

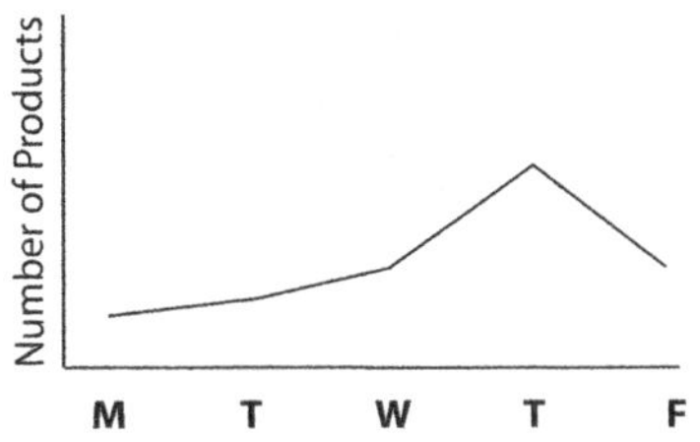


### Practice: Telling a Story

Answers will vary but should reflect the shape of the graphs.

### Extension: The Shape of a Line from Your Life

Answers will vary, but the shape of the graph should reflect the story told.

### Extension: Elevator Traffic

Answers will vary but should include the following ideas: There is a lot of "up" elevator traffic in the morning, especially between 8:30 and 9:00; then the "up" traffic significantly decreases. Around 12:30, the "up" traffic increases, probably due to staggered lunch times. Traffic going down starts to pick up around 12:00 until about 2:00 with people probably going to lunch. It appears that most people leave work between 5:00 and 5:30, which shows a sharp increase in "down" traffic.

### Test Practice

1. (1)
2. (2)
3. (1)
4. (3)
5. (3)
6. 4

## Lesson 6: Roller-Coaster Rides

### Activity 1: Amusement Park Attendance

1. Answers will vary but should include ideas such as the following: Overall, attendance has increased over a nearly 10-year period, from about 255 million to 310 million. There was only one year of decline—from '13 to '14. From '17 to '18 showed no increase or decline. In all other years, there was an increase in attendance.
2. Answers will vary.
3. Answers will vary.

### Activity 2: (Station 1 and 2) Ups and Downs

Answers will vary, but sketch of "Pharmacyclics Stock" should include the following: an upward slant from September to December, then a steep downward slant from December to April, with a less steep slant from April to June.

Answers will vary, but sketch of "Average Milk Prices" should include the following: not much change between 2012-2014 followed by an increase in 2015 and a fairly sharp decline between 2015 to 2018. The sketch then takes a sharp upward slant beginning in 2019.

### Activity 2: (Station 3) Jerri's Jobs

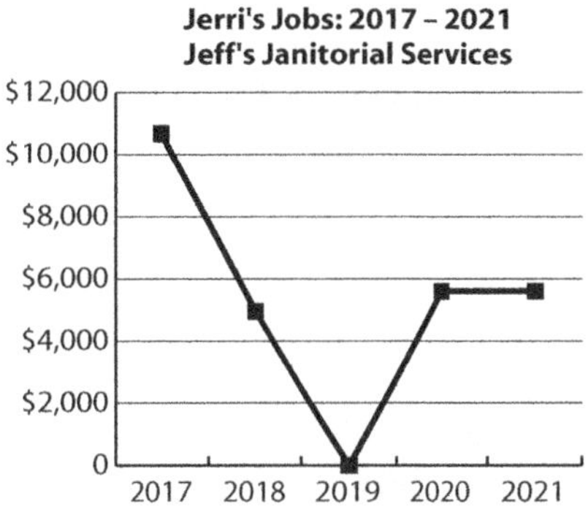


1. The greatest change occurred from **2017 to 2018**.
2. Jerri's earnings decreased by half in **2018**.
3. Answers will vary but should include ideas such as the following: Jerri's earnings steadily decreased from 2017 to 2019. Jerri didn't earn any money from Jeff's Janitorial Services during 2019. Jerri earned the same amount for 2021 as she did for 2020.
4. Answers will vary but should include ideas such as the following: Jerri's earnings in 2021 were almost half what she made in 2017.

## Activity 2: (Station 4) The Price of a Subway Ride

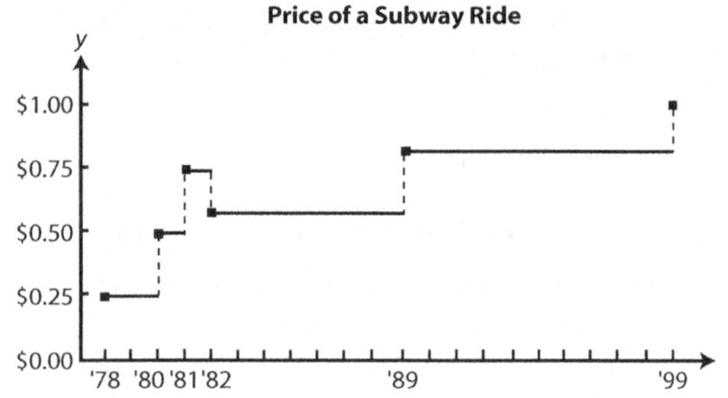


1. The longest time period with no change in price is from 1989 to 1998.
2. Answers will vary. Sample statement: A connected line gives the impression that there was a gradual change in the cost of subway fares.

## Practice: National Park Visitors

1. Answers will vary but should include ideas such as the following: After an increase in visitors from 1996 to 1998, there seemed to be a shift in interest in National Parks; in 1999–2002, there was a decrease in the number of visitors.
2. Answers will vary.
3. Answers will vary.

## Practice: Salary History

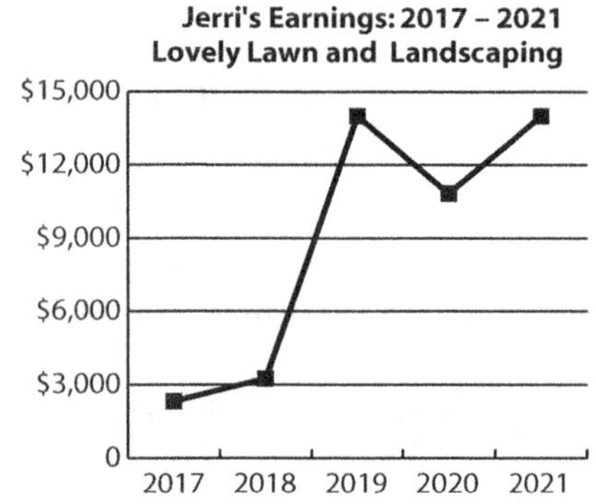


1. The greatest change in Jerri's compensation occurred between 2018 and 2019.
2. Her earnings more than doubled in **2019**.
3. In **2020**, Jerri's earning decreased by **$3,162.78**.
4. Answers will vary but should include ideas such as the following: Jerri's earnings increased over $11,500 between 2017 and 2021. Jerri earned six times as much in 2021 as she did in 2017.
5. Answers will vary. Sample story: Jerri worked only summers during her first two years at Lovely Lawn and Landscaping. In 2019, she was able to work all season for the company since she bought a snowplow for winter snow removal. In 2020, there was little snow to remove, so her income during the winter months dropped. In 2021, there was plenty of snow, which meant income for Jerri.

## Practice: More Practice with Intervals

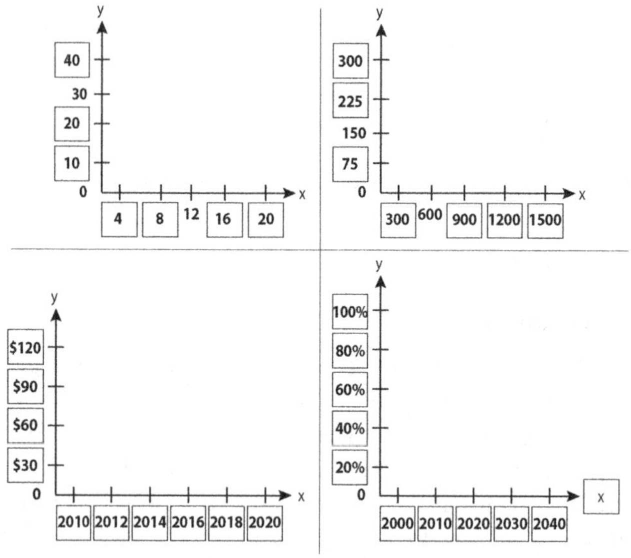


## Extension: Dropping Below Zero

1. At the ballpark, a hot-dog vendor graphed his sales for one summer. In May, he lost $**500** because he had more hot dogs on hand than he could sell. In June, the ballpark was sold out, and he made $**2,500**. In July, the profits stayed steady at $**2,500**. In August, he broke even, with **zero** profit and zero loss. In September, he had a loss of $**250**.
2.

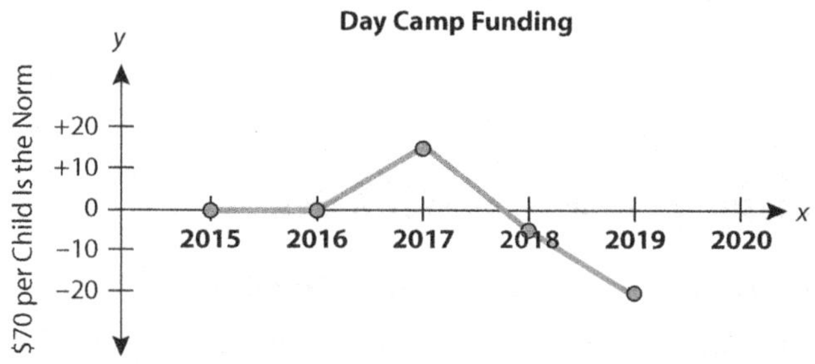

**Test Practice**

1. (2)
2. (4)
3. (5)
4. (4)
5. (3)
6. 60 million

## Lesson 7: A Mean Idea

**Activity 1: The TV Is On**

1. See *Activity 1*, p. 93, for suggestions on what to look for as students find the average using Post-it Notes or pennies.
2. See *Looking Closely*, p. 96, for suggestions on what to look for.
3. 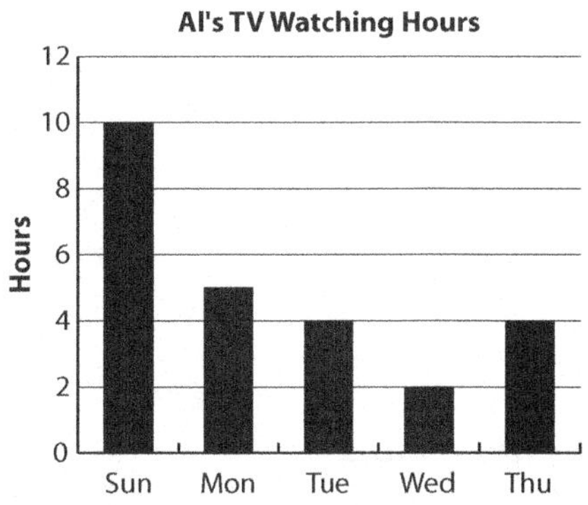

4. Jess watched TV for **five** days. The fewest number of hours she watched in a single day was **two**. The highest number of hours she watched in one day was **five**. The mean number of hours per day of watching TV for Jess was **three**. Al watched TV for **five** days. The fewest number of hours he watched in a single day was **two**. The highest number of hours he watched in one day was **ten**. The mean number of hours per day of watching TV for Al was **five**.

**Activity 2: A Mean Score**

**Round 1**

Predictions will vary but should be reasonable estimates.

Mean score Polly: exact answer: **86.8**; rounded answer: **87**; visually estimated: **86–87**.

Mean score Malaika: exact answer: **91.5**; rounded answer: **92**; visually estimated: **91–92**.

Mean score Denise: **85**

Notes about the data set will vary but should reflect the information in the data sets.

**Round 2**

Predictions will vary but should be reasonable estimates.

Actual

Polly: Final exam grade: **80** Final average: **86**

Malaika: Final exam grade: **71** Final average: **76**

Denise: Final exam grade: **82** Final average: **93**

**Round 3**

Predictions will vary but should be reasonable estimates.

Mean score Warner: **436**

Mean score Haima: **422**

Notes about the data set will vary but should reflect the information in the data sets.

**Round 4**

Answers will vary but must reflect a total score of 2,250. Sample answer: Warner needs 480 points on his Math test, which is 70 points higher than his present score to pass with a total of 2,250 points.

Answers will vary but must reflect an actual total. Sample answer: Haima retook her Science test and scored 480; she scored 450 on her Lit test and 470 on her Math test. Her new total is 2,280 points for a passing score.

**Practice: Mental Mean**

1. Mean age: **7.5**
   Mean ticket price: **$7.50**
   Mean number of miles traveled: **75**
   Mean number of students: **750**
   Mean amount of scholarship: **$7,500**
   Mean bid: **$7.5 million**
2. Mean age: **57**
   Mean number of miles traveled: **570**
   Mean number of students: **5,700**
   Mean amount of scholarship: **$57,000**
   Mean bid: **$5.7 million**
3. **Set 1**
   Ages: **6**
   Lottery winnings: **$60**
   Miles: **600**
   Bids: **$6,000**

**Set 2**

Ages: **27.5**

Lottery winnings: **$275**

Miles: **2750**

Bids: **$27,500**

**Practice: Among Friends**

1. a. The average of everyone's pocket money is **$5.90**.
   b. **Yes**, 7 bagels and coffee cost $15.75. One person alone has $20.00.
   c. **No**, 8 movie tickets cost $56.00. They only have a total of $41.30.
   d. To find the average, Ella totaled and divided by **7**.
2. The mean height for these four-year-olds is **40.375** (rounded answer: **40**; visually estimated: **40–41**).

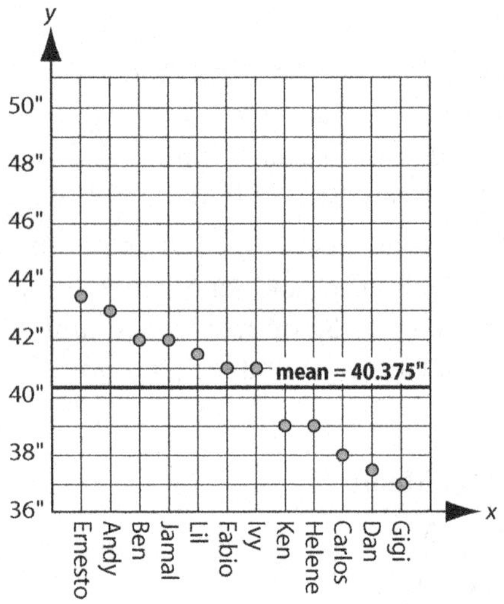


   a. Carlos's brothers call him a shrimp, but in fact he is only **2.375** (rounded answer: **2**) **inches** shorter than the mean height for four-year-olds.
   b. Ivy's parents call her a big girl. At four, she is **.625** (rounded answer: **1**) **inches** taller than the mean height for children her age.

**Practice: Data Collection**

Answers will vary.

**Extension: The Nurse's Mean**

1. a. Answers will vary but should include statements similar to the following: Mari's busiest day of the week is Friday, when she sees an average of 35 children. Her slowest day of the week is Wednesday. She saw a total of 40 children on Wednesdays over a month, for an average of 10 per Wednesday.
   b. Monday average: **18**
      Tuesday average: **15**
      Wednesday average: **10**
      Thursday average: **29.5**
      Friday average: **35**
   c. If Mari needed to attend a meeting, **Wednesday** is the day least likely to interfere with her work in school.
   d. To defend her job, she should report **Week 2** in which she saw 135 children.
   e. Answers will vary.
2. Answers will vary.

**Test Practice**

1. (1)
2. (2)
3. (1)
4. (2)
5. (1)
6. 2.5

## Lesson 8: Mystery Cities

**Activity 1: Reading Climate Graphs**

1. Answers will vary but should include ideas such as the following: Both graphs display two types of information—temperature and precipitation. Both graphs display climate over a period of a year. Both graphs use two different $x$-axes in order to depict two different types of measurement—°F and inches of precipitation.
2. Answers will vary but should include ideas such as the following: Even though the patterns of lines and bars look similar, they represent different amounts of precipitation because the scales for inches are different.
3. **Climate Graph 1**
   Temperature: **22 °F**
   Precipitation: **5 inches**
   **Climate Graph 2**
   Temperature: **24 °F**
   Precipitation: **1.5 inches**

### Activity 2: Mystery Cities

City 1 matches Graph 4—Phoenix, AZ

City 2 matches Graph 3—Miami, FL

City 3 matches Graph 1—San Diego, CA

City 4 matches Graph 2—Boston, MA

City 5 matches Graph 5—Seattle, WA

### Practice: Incomplete Climate Graphs!

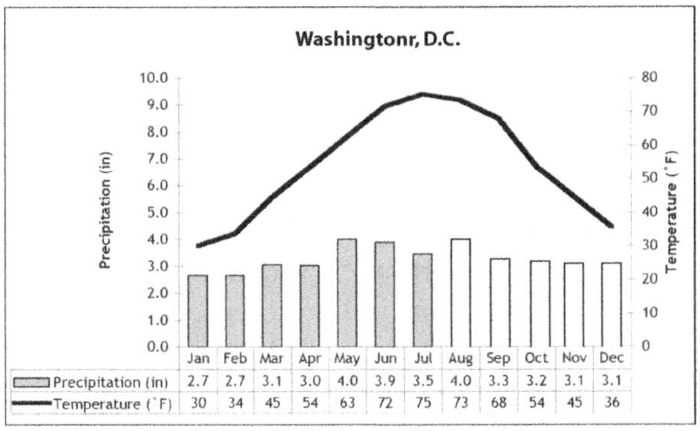


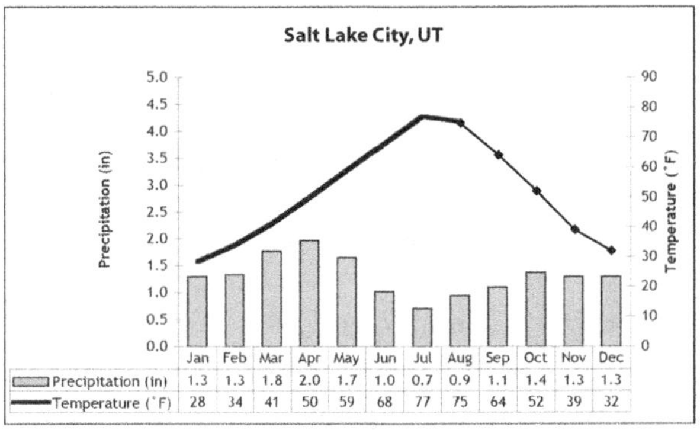


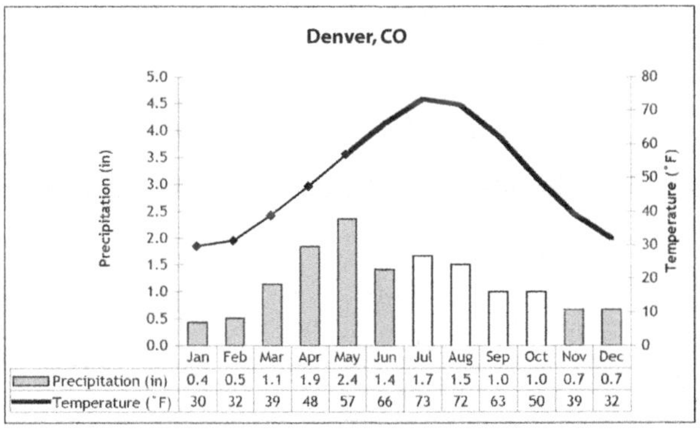


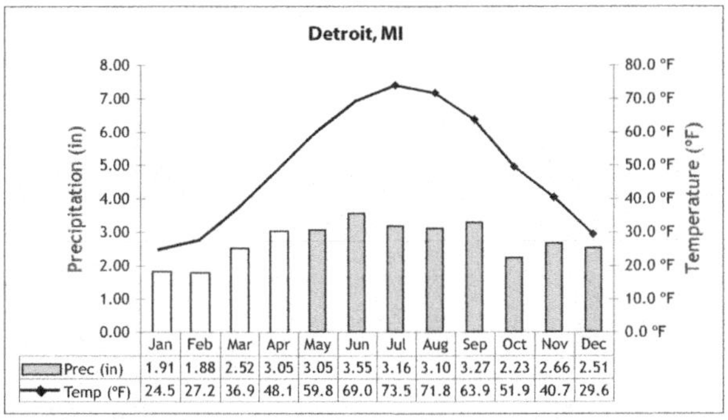


### Extension: Quality of Life 1

| Month | Utility Cost | Cost for Apt. 1 | Cost for Apt. 2 |
|---|---|---|---|
| Jan | $175 | $2,825 | $2,800 |
| Feb | $172 | $2,822 | $2,800 |
| Mar | $160 | $2,810 | $2,800 |
| Apr | $160 | $2,810 | $2,800 |
| May | $150 | $2,800 | $2,800 |
| Jun | $170 | $2,820 | $2,800 |

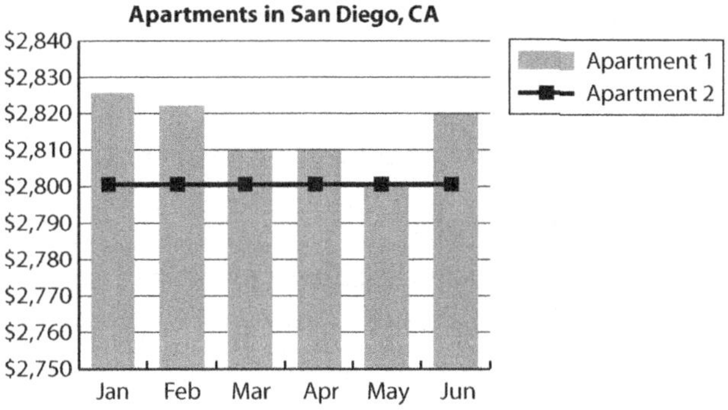


### Extension: Quality of Life 2

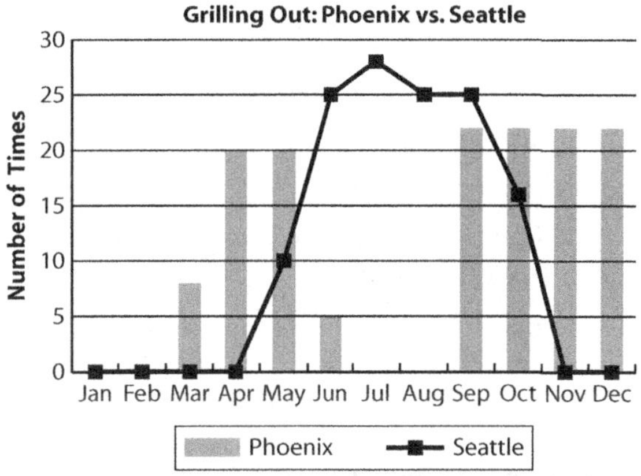


Answers will vary but should include ideas such as the following: According to the graph, he did not grill out *all the time*; he only grilled out 8 months out of the year (141 times) while in Phoenix. In Seattle, he grilled out 6 months out of the year (129 times).

Answers will vary but should include ideas such as the following: When he lived in Phoenix, he grilled out a great deal in April and May and then again from September through December because these months weren't too cold (they were above 50°F) or too hot (they were 80°F and below). In Seattle, those same temperature ranges (50°F to 70°F) occurred between the months of April and October.

### Test Practice

1. (2)
2. (2)
3. (5)
4. (1)
5. (3)
6. 1999

## Lesson 9: Median

### Activity 1: Line Up

Answers will vary depending on students in class.

### Activity 2: At the Party

Answers will vary. Sample answer for Story 1: ages of kids: 3, 3, 5, 6, 8; ages of moms: 24, 24, 25, 25, 28

Answers will vary. Sample answer for Story 2: ages: 14, 15, 15, 16, 17, 18, 20, 24, 27

Answers will vary. Sample answer for Story 3: ages: 1, 1, 1, 2, 34, 40, 51, 56

Answers will vary. Sample answer for Story 4: ages: 48, 56, 71, 71, 75, 78, 80, 81, 83

### Practice: Finding the Median

1. The median score is **80**.
2. The median price of a dozen large Grade A eggs in 1980 was **$0.84**.
3. The median height of this group is **64.5 inches.** The height 64.5 inches is the same as 5-3/8 feet, which is very close to 5-1/2 feet.

### Extension: The Average Wage

1. The median weekly wage for the staff at M&M Inn is **$750.**
2. The mean weekly wage for the staff at M&M Inn is **$855**.
3. The mean represents all of the staff's salary divided by the number of people who work at M&M Inn. The mean of $855 is higher than the median because it includes the high salary of the owner.
4. The median shows that there are the same number of staff members who earn less than $750 as there are who earn more than $750.
5. The baker would want to use the mean salary of $855 to argue his point, since his salary is $105 less than the mean. His salary is the same as the median; half the staff makes the same or less than he does so the median is not a good arguing point if he wants a raise.

### Extension: Mean or Median?

1. The median is **$775**. The mean is **$819.44**. Answers will vary. Sample answer: The median best describes the center of the data because there are an equal number of fees above and below $775. There are 5 fees lower than the mean of $819.44 and only 4 higher than that.
2. The median is **22.5**. The mean is **31.4**. Answers will vary. Sample answer: The median best describes the center of the data because there are an equal number of phone calls above and below 22.5 minutes. If the outliers were dropped (3, 5, and 120 minutes—2 hours), there would be only 2 calls longer than 31.4 minutes but 5 less than 31.4 minutes.
3. The median is **14**. The mean is **18.55.** Answers will vary. Sample answer: The median best describes the center of the data because there are 5 runners who run more than 18.55 miles per week but 6 who run less than 18.55; on the other hand, there are also 5 runners who run more than 14 miles per week and only 4 who run less than 14 miles (there are 2 who run exactly 14 miles per week).
4. The median is **$1,182.50**. The mean is **$1,275**. Answers will vary. Sample answer: The median best describes the center of the data because there are an equal number of truck drivers earning above and below $1,182.50. There are 4 drivers earning less than the mean of $1,275 and 2 more than that.

### Test Practice

1. (5)
2. (2)
3. (4)
4. (4)
5. (3)
6. 42 days

## Lesson 10: Stock Prices

### Activity 1: Three Companies

Answers will vary but should be close to the estimates of the examples below:

### Financial Co.

| Day | Price |
|---|---|
| 1 | $17.80 |
| 2 | $18.20 |
| 3 | $17.75 |
| 4 | $17.00 |
| 5 | $18.05 |
| 6 | $18.05 |

Mean: **$17.81**

Median: **$17.93** (exact answer: $17.925)

### Energy Enterprises

| Day | Price |
|---|---|
| 1 | $32.00 |
| 2 | $29.00 |
| 3 | $44.00 |
| 4 | $44.00 |
| 5 | $44.00 |
| 6 | $44.00 |

Mean: **$39.50**

Median: **$44.00**

### Technology, Inc.

| Day | Price |
|---|---|
| 1 | $16.40 |
| 2 | $17.30 |
| 3 | $16.10 |
| 4 | $16.30 |
| 5 | $18.00 |
| 6 | $18.80 |

Mean: **$17.15**

Median: **$16.85**

### Activity 2: Altering Scale

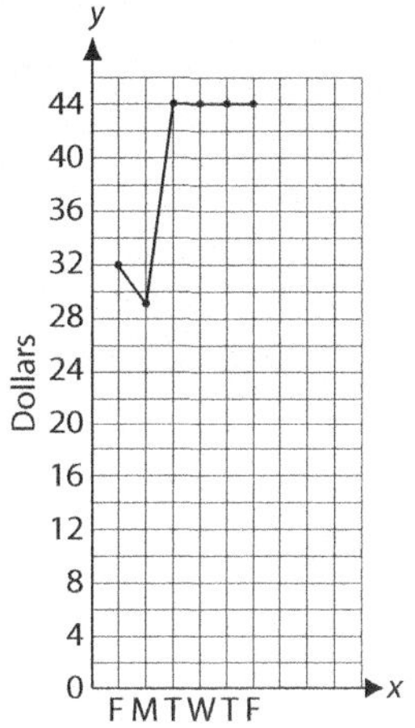


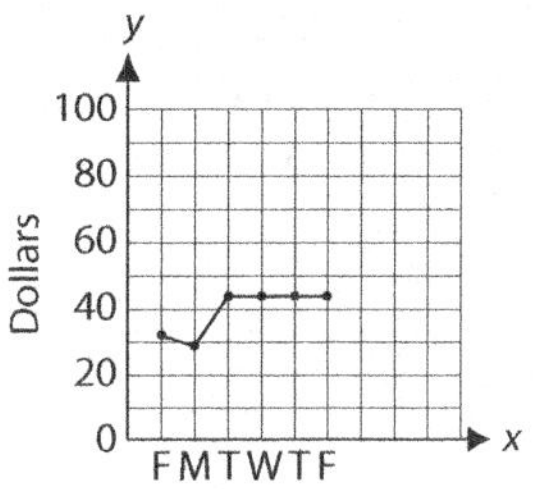


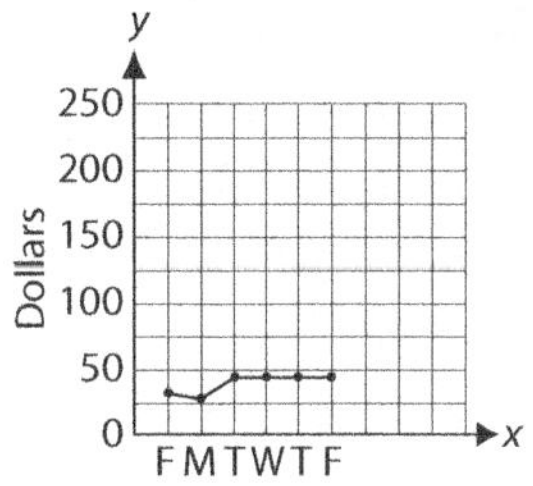


1. Answers will vary.
2. Answers will vary.
3. Answers will vary but should include a statement similar to the following: The smaller the intervals of the scale, the steeper the lines.
4. Answers will vary. Sample answer: When the scale is $2.00, the line is **steep**.
5. Answers will vary. Sample answer: When the scale is $10.00, the line is **not as steep**.
6. Answers will vary. Sample answer: When the scale is $25.00, the line is **fairly flat**.
7. Answers will vary. Sample answer: The smaller the interval, the **steeper the line.**
8. Answers will vary. Sample answer: The larger the interval, the **flatter the line**.

### Practice: More Stocks—From Graphs to Tables

### Wal-Mart

| Day | Price |
|---|---|
| 1 | $45.50 |
| 2 | $47.30 |
| 3 | $48.30 |
| 4 | $49.25 |
| 5 | $49.25 |

High for the week: **$49.25** Low for the week: **$45.50**

**Microsoft**

| Day | Price |
|---|---|
| 1 | $44.00 |
| 2 | $45.75 |
| 3 | $47.00 |
| 4 | $49.00 |
| 5 | $48.20 |

Opening price for the week: **$44.00**
Closing price for the week: **$48.20**

**McDonald's**

| Day | Price |
|---|---|
| 1 | $23.80 |
| 2 | $23.00 |
| 3 | $23.40 |
| 4 | $22.50 |
| 5 | $22.75 |

Mean: **$23.09**

Median: **$23.00**

**Practice: From Graphs to Stories—Stock Prices**

**Graph 1.** Answers will vary. Sample stories may be similar to the following:

Stock prices for Wal-Mart began the week at $45.50. Prices rose steadily through Thursday, with the sharpest increase between Monday and Tuesday. Friday's prices showed no change from Thursday's, peaking at $49.25.

**Graph 2.** Answers will vary. Sample stories may be similar to the following:

Stock prices for Microsoft began the week at $44.00. Prices rose steadily through Thursday, peaking at $49.00. Friday's prices showed a downhill trend, with the price of stock ending the week at just over $48.00.

**Graph 3.** Answers will vary. Sample stories may be similar to the following:

It was a rocky week for stock prices for McDonald's, with prices fluctuating from Monday's high of $23.80 to Thursday's low of $22.50. Friday's prices rebounded slightly, ending the week with a price of $22.75.

Answers will vary, but should be based on the information in the three graphs.

**Extension: What Is New with the Stock Market?**

Answers will vary.

**Test Practice**

1. (4)
2. (4)
3. (3)
4. (2)
5. (3)
6. 9.75

## Closing the Unit: Stock Picks

**Activity 1: Stock Picks**

Answers will vary. See *Looking Closely*, p. 136, for suggestions on what to look for in student work. Sample set of answers based on choosing two companies—Hershey Food Corporation and Heinz:

1. Company 1 Hershey (HSY)

   Company 2 Heinz (HNZ)
2. At the beginning of January 1999, Hershey stock prices were about $56.00. After a quarter of upward movement, Hershey stock fell dramatically and hit an all-time low of $48.00 in October 1999. In year two, the stock price rose after a first quarter stall with no price increase. Year 3 and 4 saw a series of ups and downs for the stock. In year 5, the stock hit an all-time high of about $77.00 before declining at the end of the quarter. In the five years reported, the stock prices increased from $56.00 to $75.00.

   Heinz began the first year at $43.00 and increased prices to $45.00 the second quarter, its highest price over the five years, before plummeting. Year 2 stock prices were $31.00. Year 2 showed steep ups and downs; year 3 showed ups and downs, but not nearly as dramatic as year 2. Year 4 also had its ups and downs. Year 5 began at the stock's lowest price of $29.00. Year 5 did show improvement until the last quarter, when it declined to an ending price of $35.00. In the five years reported, the stock prices decreased from $43.00 to $35.00, a loss of $8.00.

   Hershey began at $56.00, while Heinz began 1999 at $43.00, a difference of $13.00. By January 2004, Hershey had increased to $75.00, while Heinz had dropped to $35.00, a difference of $40.00.

3. Hershey's stock prices showed the greatest change between July and October 2000, when its prices rose from $54.00 to $65.00.

   For Heinz, the greatest change also occurred between July and October 2000, when stock prices increased from $34.00 to $43.00.

4.

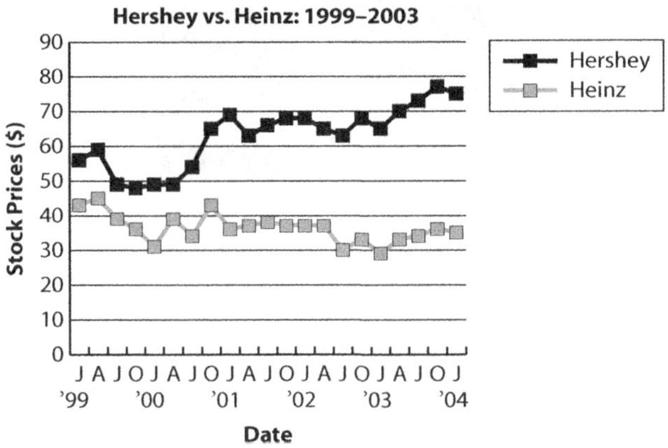


   Answers will vary. Sample answer: By making sure both graphs have the same scale, it looks like the two have followed a similar trend, although Hershey's stock prices are consistently higher than Heinz's.

### Activity 2: Review Session

Answers will vary.

### Activity 3: Final Assessment

Refer to the *Final Assessment Checklist*, p. 164, and *Looking Closely*, p. 136, for suggestions when observing students.

### Task 1: Decisions

1. Answers will vary: Sample answer: Where to take my nephews when they come here for a visit.
2. Answers will vary: Sample answer based on above sample: Important criteria to consider are my nephews' interests, the cost involved, and the travel time needed.
3. Answers will vary: Sample answer based on above sample: I will need to contact my nephews and ask them what they like to do. I will check online to find out the costs of places of interest in the area. I will check a map to see about how far it is to drive to the various places of interest.
4. Answers will vary: Sample answer based on above sample: I'll make a table with three columns once I figure out what places might interest my nephews:

| Place of Interest | Estimated Cost Involved | Driving Time |
|---|---|---|
| | | |

### Task 2: Mind Map

Answers will vary.

### Task 3: Snowfall Data

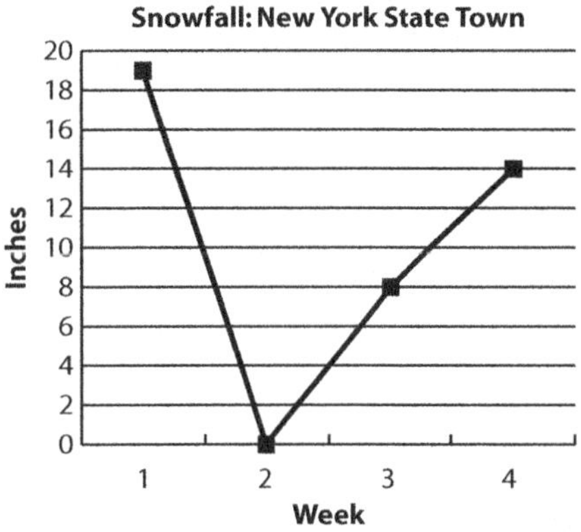


### Task 4: Tables and Line Graphs

1. Answers will vary. Sample sentences may be similar to the following: In July 2003, AT&T stock was about $21. Shortly thereafter, it sank to a low of almost $18 a share before rebounding and showing a steady increase in price.
2. Answers will vary but should be close to the estimates in the sample below:

| Month | Price of Share |
|---|---|
| July | $21.00 |
| August | $23.25 |
| September | $21.50 |
| October | $18.50 |
| November | $20.00 |
| December | $20.50 |
| January | $22.00 |

3. If you wanted to see the ups and downs more clearly, you would need a scale of **<u>$5 per interval</u>** (or more).
4. You could report the average cost of a share in several ways: mean, median, mode.

5. a. Mean price of BankOne stock: **$49.90**.
Median price of BankOne stock: **$51.00**.
   b. The difference between the mean and median is **$1.10**.
   c. Answers will vary. Sample answer: The median seems to more closely reflect the data since there are two data points on $51.00 and one just below and just above it. There is only one data point that is not very close to $51.00.

### Task 5: Printing Plant

1. Answers will vary but should include the following idea: The greatest number of rejects occurs on Fridays.
2. The mode is 5.
3. The mode indicates that 5 was the number of rejects reported more often than any other number of rejects.
4. (1) Training for workers
5. a.

**Printing Plant Errors: Machine vs. Human**

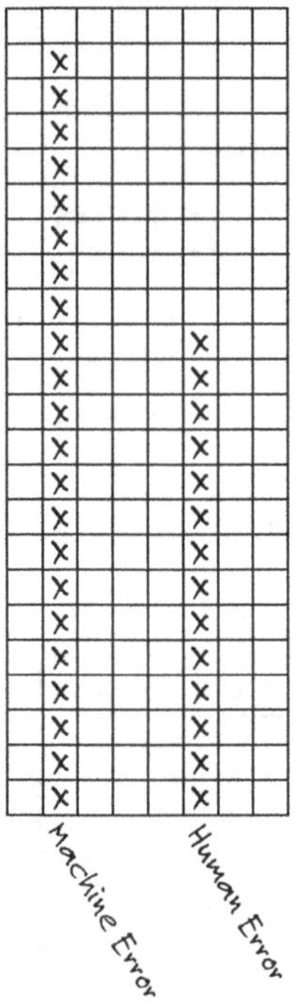


   b.

**Printing Plant Errors: Machine vs. Human**

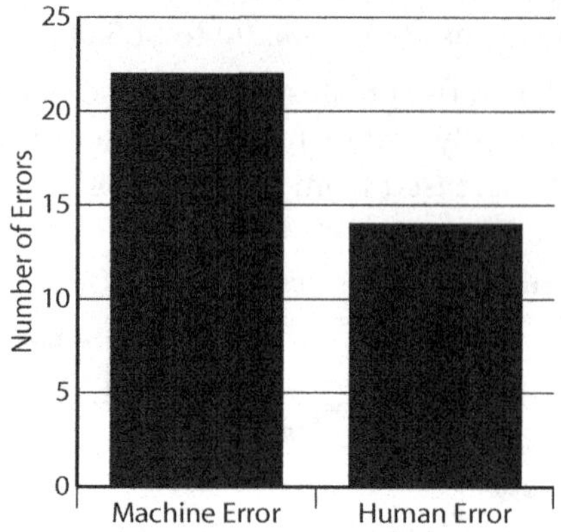


   c.

**Printing Plant Errors: Machine vs. Human**

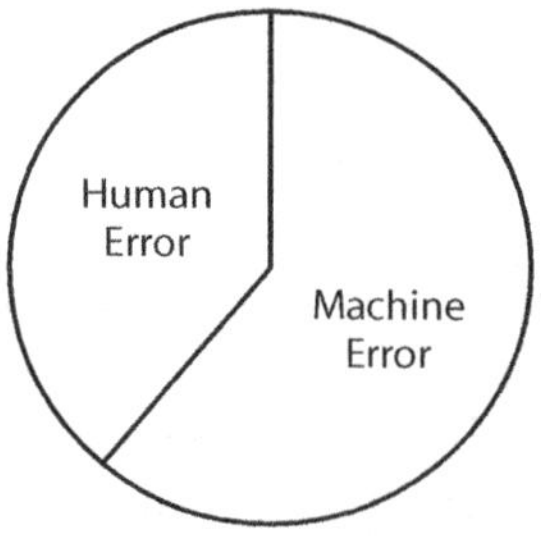


6. Answers will vary. Sample answers: The human errors represent less than half of the total errors reported. Machine errors account for a little less than two-thirds of the total errors reported during the three days.
7. Answers will vary. Sample answer: The bar graph shows very readily how the two categories compare, and it is easy to figure out exactly how many of each error were reported.

### Task 6: Story Graph Match

1. Story b matches the graph with training and cost in dollars; Story d matches the graph with participants and training.
2. Answers will vary. Sample answer: Tony's ability to cut costs and continue to provide at least 15 trainings per quarter shows his good work. Tony was able to reduce the cost of trainings while increasing the number of trainings during Quarter 1 and Quarter 2; during Quarter 3 and Quarter 4, he maintained the same cost while increasing trainings.

### Task 7: Test Type Questions

1. (5)
2. (5)
3. (4)
4. (2)

# Sources and Resources


### Mathematics Education

The National Council of Teachers of Mathematics (NCTM) publishes several excellent resources.

- *Principles and Standards for School Mathematics.* Reston, VA: NCTM, 2000.
- The NCTM journals: *Mathematics Teaching in the Middle Grades, Mathematics Teacher*, and *Journal for Research in Mathematics Education.*
- http://www.nctm.org.

Brown, Catherine A., and Edward Silver. "Data Organization and Interpretation." In *Results from the Fourth Mathematics Assessment of the National Assessment of Educational Progress*, edited by Mary M. Lindquist. Reston, VA: National Council of Teachers of Mathematics, 1989.

Fendel, D., D. Resek, L. Alper, and S. Fraser. *Interactive Mathematics Program*, Berkeley, CA: Key Curriculum Press, 1997.

Lappan, G., J. Fey, W. Fitzgerald, S. Friel, and E. Phillips. *Connected Mathematics Series*, Palo Alto, CA: Dale Seymour Publications, 1998.

Russell, S. J., et al. *The Investigations in Number, Data, and Space Curriculum.* White Plains, NY: Dale Seymour Publications, 1998.

Zawojewski, Judith, and J. Michael Shaughnessy. "Mean and Median: Are They Really So Easy?" *Mathematics Teaching in the Middle School* 5 (March 2000): 436–40.

### Mathematics and Numeracy Education for Adults

Curry, D., M.J. Schmitt, and S. Waldron. *A Framework for Adult Numeracy Standards: The Mathematical Skills and Abilities Adults Need to Be Equipped for the Future.* Boston: World Education, 1996. (This framework was developed by members of the Adult Numeracy Network.)

Massachusetts Department of Education. *Massachusetts Adult Basic Education Curriculum Frameworks for Mathematics and Numeracy.* Malden, MA: Massachusetts Department of Education, 2001.

McGee, Lynn. "Where Are Our Clothes Made? Field Testing the EMPower Curriculum." *The Math Exchange* 3 (Spring 2002): 8–12.

"Spreadsheet Guide to Using Microsoft Excel" adapted and reprinted with permission of Pat Fina, Community Learning Center, Cambridge, MA.

Stein, S. *Equipped for the Future Content Standards: What Adults Need to Know and Be Able to Do in the Twenty-First Century.* ED Pubs document EX0099P. Washington, DC: National Institute for Literacy, 2000.

### Health

For facts about blood and blood banks, see the American Association of Blood Banks, http://www.aabb.org.

For HIV/AIDS data, see http://www.aidshotline.org/crm/asp/refer/statistics.

### Immigration

For the Department of Homeland Security, see https://www.dhs.gov/immigration-statistics.

For the United Nations High Commission on Refugees, see http://www.unhcr.ch.

### Mathematics Resources Online

For lots of fun facts, see http://www.infoplease.com or http://www.factmonster.com.

Make your own graphs on http://nces.ed.gov/nceskids/graphing. The site has an easy-to-read interface.

For interesting data on crime trends and graphs with matching statements, see the Department of Justice, Bureau of Justice Statistics, http://www.ojp.usdoj.gov/bjs/glance.htm.

For interactive Java applets for students or teachers, see http://www.shodor.org.

For math questions see https://www.nctm.org/drmath/.

### Parks, Land, Recreation

"Change in Land Use, 1985–1999." *Massachusetts Audubon Losing Ground: At What Cost?* Reprinted with permission from the Massachusetts Audubon Society, 2003.

Attendance data on amusement parks reprinted with permission from the International Association of Amusement Parks and Attractions, http://www.iaapa.org.

"Ups and Downs at the Park" reprinted with permission from the *Boston Herald*, June 4, 2000.

For data on National Park Service visitors, see https://www.nps.gov/aboutus/visitation-numbers.htm.

### Stocks and Business

"Going Wireless" graphs reprinted with permission from the Chicago Tribune Company, 2000.

NASDAQ home page reprinted with permission from NASDAQ Stock Market, Inc.

Tyson, Eric. *Investing for Dummies*, 2nd edition. New York: Wiley Publishing, 1999.

### Transportation

For commuting times, see the Census Bureau, http://www.census.gov/.

"Countries with the Greatest Number of Elevators," "Elevators per 10,000 Population for the Above Countries," and "Elevator Traffic" reprinted with permission from the National Building Museum. For more information, see the exhibit catalog: *Up, Down, Across: Elevators, Escalators, and Moving Sidewalks.* London: Merrell Publishers, 2003.

### Work

For data on inflation, employment, wages, and safety, see the Department of Labor, https://www.dol.gov/.

For data on clothing manufacturing, see https://cleanclothes.org/fashions-problems.

"Changing Needs of the American Workforce" reprinted with permission of the Association for Career and Technical Education, http://www.acteonline.org.

Made in the USA
Las Vegas, NV
07 June 2026

48302805R00129